TURNING **SEWAGE** INTO **REUSABLE WATER**

WRITTEN FOR THE LAYPERSON

TURNING **SEWAGE** INTO **REUSABLE WATER**

THIS MAY BE THE NORM OF THE FUTURE

MADAN ARORA
and
JOE REICHENBERGER

Archway Publishing books may be ordered through booksellers or by contacting:

Archway Publishing
1663 Liberty Drive
Bloomington, IN 47403
www.archwaypublishing.com
1 (888) 242-5904

ISBN: 978-1-4808-1376-2 (sc)
ISBN: 978-1-4808-1378-6 (hc)
ISBN: 978-1-4808-1377-9 (e)

Library of Congress Control Number: 2015900187

Print information available on the last page.

Archway Publishing rev. date: 2/23/2015

When the well's dry,
we know the worth of water.

— Benjamin Franklin

Contents

Preface

We have spoken about wastewater treatment, also known as sewage treatment, and water reuse to various civic groups, such as Kiwanis clubs; medical and professional groups comprised of doctors and nurses, students, and teachers; council members of small cities; and others. During these talks and discussions, it became obvious that despite the wealth of knowledge and information in their own areas of expertise, they have very little information on how the wastewater generated in their homes and neighborhoods is conveyed to a remote location through a network of sewers and pumping stations, treated by seemingly complex processes, and often recycled back to the neighborhood to water their lawns and parks. All this is done through an "out of sight, out of mind" system.

Because water seems abundant in some areas, the concept of recycling this spent water even for such mundane uses as watering lawns or flushing toilets is repugnant to many. The "ick" factor is still present in many communities. People cringe at the idea that they may be using water for irrigation that originated as wastewater from a nearby neighborhood or community. The very idea that someone else's wastewater may be a part of their drinking water is even more disconcerting.

But in most areas, fresh, drinkable water is a limited resource and precious commodity essential to ensuring a sustainable community. As water becomes more scarce and expensive, decisions will be made on water recycling. We believe the general public and key community leaders should have sufficient technical understanding to make correct decisions regarding water and wastewater systems and recycling, many of which involve the expenditures of large sums of money or long-term energy commitments. Many of these decisions are influenced by the ick factor.

We became convinced there is a need for an easy-to-understand book tailored to the general public, laypeople, and nonwastewater professionals to explain what happens to the water they use and "throw away" and how it can be recycled safely and efficiently. As communities grow from rural to more urbanized, a decision needs to be made: Do we install sewers or continue to rely on our individual systems? What are the alternatives?

We realize there are a number of books written for people who have academic training and/or experience in the field of wastewater treatment and recycling. But these books, even though well written, are filled with theories, equations, coefficients, complicated diagrams, terminologies, and concepts. They serve the needs of a very limited group: those who have careers in this important field.

Few, if any, books explain the technology of wastewater treatment, collection, and reuse in an easy-to-read and understandable language for laypeople and nonwastewater professionals such as science and urban planning students; parents and students of high school and middle school age; journalists writing about the latest bond measure affecting their community; city council and sanitary board members weighing decisions on their wastewater system; and other influential community leaders.

We believe there is a need for a book written for those with no training in this field but a keen desire to know what happens to the wastewater generated in their homes and communities. This book was written to fill that void. We hope you will find it useful and will begin to appreciate that the water supplies on our planet are not unlimited and must therefore be conserved, used, treated, and recycled. We hope that you will appreciate the science behind the conveyance of water to your home and neighborhood, the collection of wastewater from your home and neighborhood, the complexities of treatment of wastewater and the issues surrounding it, and the safety of recycled water for reuse.

As a final thought, we also hope that parents, teachers, and counselors who advise and assist students in their career choices will suggest the water and wastewater field as a career to those who like math and the biological and physical sciences, who want to do things personally rewarding, and who want a challenge. These careers can be in the regulatory area, in system and facility design and construction, and in system and facility operation. Between the two of us, we have over one hundred years of experience in this important, ever-changing, challenging field, and we have never once looked back.

We acknowledge with gratitude the assistance of Vamsi Seeta and Leonor DeGuchy in helping to edit and finalize the manuscript in accordance with publisher's guidelines. Without their help, this book could not have been completed in its present form in a timely manner.

Madan Arora | Joe Reichenberger

Chapter 1

Introduction

In developed countries, virtually all people have safe drinking water piped into their homes for food preparation; cooking; dish, hand, and clothes washing; bathing; and other sanitary purposes. The water is used and then sent down a drain, leading to a sewer that conveys it to a remote location for treatment and disposal or reuse. Except for civil and environmental engineers (called sanitary engineer in the middle of the twentieth century) and those who are responsible for the treatment and disposal of used water (wastewater or sewage), most people do not think twice about the water once it is used and flushed or drained away. "Out of sight, out of mind." This book explains wastewater collection treatment, disposal, and reuse in easy-to-understand terms for those who are not wastewater professionals.

This book will be valuable to elected city, county, and state officials of jurisdictions that have wastewater collection and treatment systems and who are responsible for budgeting for the maintenance, operation, and replacement of these systems. The investment made by municipalities and local governments in wastewater collection, treatment, and disposal is enormous. The American Society of Civil Engineers (ASCE), in their 2013 Infrastructure Report Card, stated there are between 700,000 and 800,000 miles of public sewer mains in nearly 20,000 wastewater pipe systems, with nearly 15,000 treatment facilities in the United States alone. Capital needs to expand and upgrade these wastewater and storm-water systems to meet more stringent requirements will require the investment of nearly $300 billion over the next twenty years in the United States alone. Pipes represent the largest capital need, comprising three-quarters of total needs (ASCE 2013).

Understanding how these facilities work and what can be expected of them in terms of the level of treatment and the options available is invaluable in community leaders' decision-making process. Journalists reporting on community needs, costs, and consumer impacts will be able to produce meaningful and informative articles. General science and environmental science students and teachers at all levels should find this book useful and informative in making or advising on career choices. The general public will also have a better understanding of what they are paying and why those expenditures are needed.

Early History

Sewer systems are not some modern invention; sewers dating to the seventh century BC have been uncovered in Nineveh and Babylon. The Romans built extensive water and sewer systems, which were primarily for rainwater and storm water drainage, but also collected sanitary wastewater. Many of the aqueducts used to import water from remote locations stand today as a testimony to Roman engineering. The science of sewer systems did not advance during the Middle Ages; instead, masonry cesspools were constructed below living quarters. Modern sewage systems were constructed in

Europe, generally discharging into the nearest river or watercourse. Because of this, cholera outbreaks were common. Although John Harrington invented the flush toilet in the late 1500s, it was almost three hundred years before Thomas Crapper perfected it (Foil, Cerwick and White 1993).

In colonial America, pit privies were the typical method for disposing of sanitary waste. They frequently had outlets at ground level, which allowed overflow water to discharge to a yard or gutter. Eventually, as populations grew, this form of wastewater disposal created nuisance conditions. Vaults were constructed beneath the privies to allow the contents to be stored and then periodically "picked up" and disposed of. In some cases, cesspools consisting of large, lined holes were constructed to allow the wastewater to soak into the ground (Burian, et al. 2000).

George Washington's "Necessary"
at Mt. Vernon, VA
Photo by Lance Reichenberger

Dry or "pail" systems were also used in residential wastewater management in the 1800s in the United States. Pails or drawers were placed under the seats of the privies to collect human solid waste. The waste was removed periodically and disposed of close to the residence (Burian, et al. 2000). For example, George Washington, our nation's first president, had a "necessary" at his home at Mt. Vernon, Virginia.

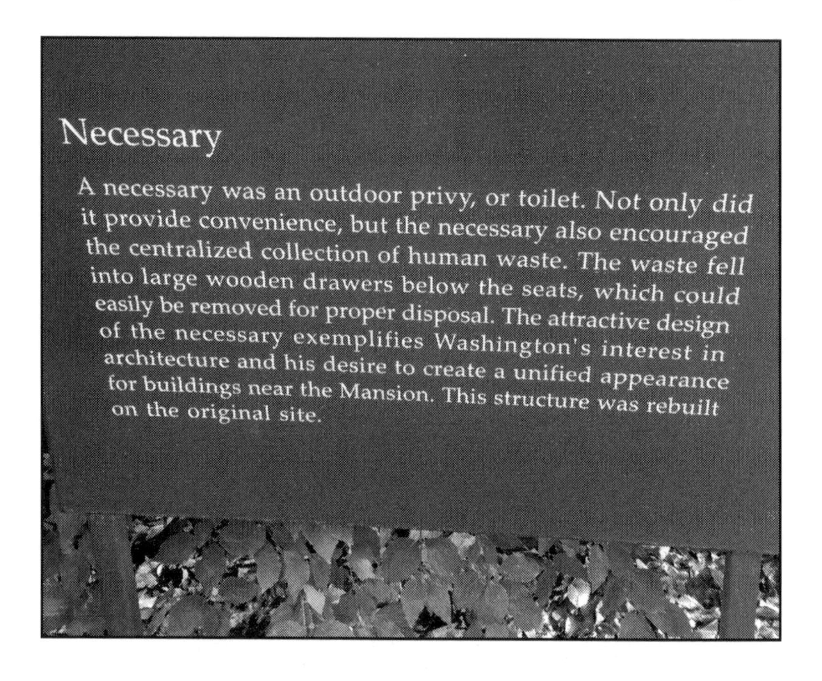

It really was not until the mid-1800s, with the discovery that germs cause many diseases, that sanitation became important. The first modern-day water-carriage sewer system was constructed in Hamburg, Germany, in 1843. By the end of the nineteenth century, most of the major cities in the United States had some form of sewer system. Typically it was combination of a storm-water runoff and sanitary system, because both had to be accommodated in the more urbanized areas. Some less-urbanized areas had just a sanitary sewer system, since there was less of a need for storm-water management. Even as late as 1924, over 88 percent of the population in cities with more than 100,000 people continued to discharge their wastewater directly to rivers, lakes, and watercourses without treatment.

Gradually, urbanizing areas were developing separate sanitary sewers to carry just the domestic wastewater—no storm water. Cost-effective treatment technologies were developing. Industrial expansion after World War II introduced toxic wastewater into wastewater systems and streams. The advent of laundry machines, dishwashers, garbage disposals, and the like increased the volume of wastewater that needed to be managed (Burian, et al. 2000). (See chapter 5 for a brief but more detailed history of wastewater collection and treatment from the earliest recorded history to the present.)

Federal Involvement

The federal government became involved in wastewater management in the United States with the promulgation of the Water Pollution Control Act of 1948, which authorized the Surgeon General, in cooperation with other federal, state, and local entities, to prepare comprehensive programs for eliminating or reducing the pollution of interstate waters and tributaries and improving the sanitary condition of surface and underground waters. Although amended several times thereafter, water quality continued to worsen through the 1960s, and a different approach was needed.

The 1960s saw the rise of environmental awareness in the United States; Rachel Carson's *Silent Spring*, published in 1962, is often cited as launching the environmental movement (Griswold 2012). The Cuyahoga River, accidently set ablaze in June 1969, was a major impetus to drive Congress to improve the quality of the nation's waters. In 2009, the Cuyahoga is now teeming with fish and aquatic life (Scott 2009).

In response to worsening water pollution, Congress enacted the federal Water Pollution Control Act Amendments of 1972, commonly called the Clean Water Act. This act had the objective of restoring and maintaining the chemical, physical, and biological quality of the nation's waters, with a goal of fishable, swimmable rivers by 1983 (33 U.S.C. 2002). (This is discussed in more detail in chapter 5.)

Wastewater Management in Rural Areas

In rural areas and suburbia, individual on-site wastewater systems are the norm. These systems typically consist of a septic tank and a leach field. A septic tank is a large tank constructed underground in the front or back yard of a residence to collect wastewater from all the plumbing fixtures. Solids settle out in the tank and undergo digestion; the liquid flows out to a perforated drainage pipe, set in a gravel bed and buried a few feet beneath the lawn, where the settled water percolates into the ground. These systems generally perform well in areas where the density of development is low, the soil allows percolation, and groundwater or rock is not close to the surface. Periodically the solids in them do need to be cleaned out. County or state ordinances regulate the installation of these systems.

Onsite Wastewater System in Rural Areas

When these areas become more urbanized, public health and water-quality concerns grow, and there is a need to replace these systems with a conventional or alternative sewer system. This is expensive and costly to the homeowner and is frequently met with substantial public resistance.

What's Next?

The following chapters describe in detail how wastewater systems work; provide some general information on cost to construct and operate these systems; and tell how we can make these systems more sustainable through energy recovery, water recycling, and biosolids use. The book ends with a brief outlook for the future.

References

33 U.S.C., 1251 et. seq. 2002. "Federal Water Pollution Control Act [As Amended Through P.L. 107–303, November 27, 2002]." November 27.

ASCE. 2013. *2013 Report Card for America's Infrastructure.* Accessed November 2, 2013.

Burian, Steven J., Stephan J. Nix, Robert E. Pitt, and S. Rocky Durrans. 2000. "Urban Wastewater Management in the United States: Past Present and Future." *Journal of Urban Technology* 7 (3): 33-62.

Foil, J. L., J. A. Cerwick, and J. E. White. 1993. "Collection Systems Past and Present, A Historical Perspective of Design, Operation and Maintenance." *Operations Forum*, December.

Godfrey Hoffman Associates. 2012. *Importance of Proper Septic Tank Design in CT.* December 29. Accessed September 20, 2014. http://www.godfreyhoffman.com/Civil-Engineering-Blog/bid/251009/The-Importance-of-a-Proper-Septic-Design-in-CT.

Griswold, Eliza. 2012. "How 'Silent Spring' Ignited the Environmental Movement." *New York Times Magazine.* Edited by Sheila Glaser. New York, September 21.

Scott, Michael. 2009. "Cuyahoga River fire 40 years ago ignited an ongoing cleanup campaign." *The Plain Dealer.* Cleveland, June 22.

Chapter 2

The Water Cycle

This chapter presents two important points: (1) good-quality water is not as abundant as many of us are inclined to believe, and (2) water goes through a full cycle from nature to us and from us back to nature.

Water Is Not Abundant

Figure 2.1 presents the distribution of water on the Earth. It is interesting, and perhaps surprising, to note that about 71 percent of the Earth's surface is covered by water, and of that, approximately 97 percent is oceans and seas. This water is mostly highly saline and unusable as drinking water unless treated using very expensive technologies—both in capital cost and recurring O&M cost. In addition, these technologies (called desalination) are feasible only in coastal communities adjacent to the ocean. Of the remaining 3 percent of the Earth's water, about 69 percent consists of polar ice and glaciers and 30 percent is underground (that is, below the surface as groundwater, not all of which can be used without some form of treatment).

That leaves only about 1 percent in rivers, streams, lakes and the atmosphere. Let's do the math. If there were one million gallons of water on and surrounding our planet, the water in fresh water lakes and rivers would be about 300 gallons, and underground would be about 8,000 gallons. So we can easily see that the water available to us is not abundant.

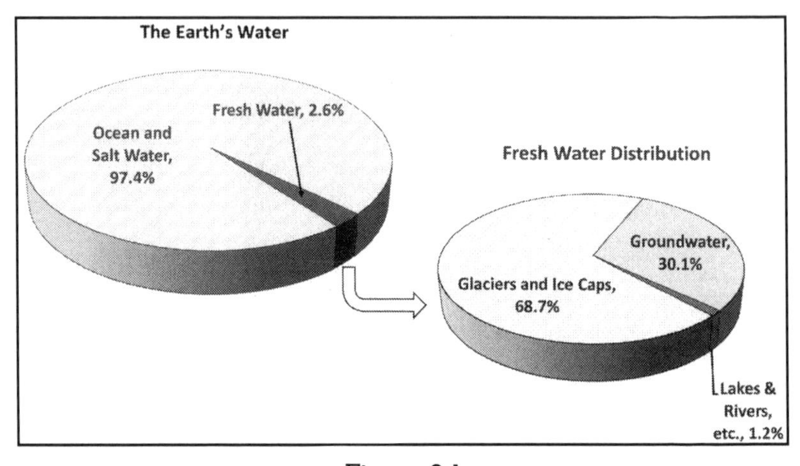

Figure 2.1
Earth's Water Distribution

Another problem that we all face is that the water is not only limited in quantity but also not increasing with time—unlike the world population, which is expected to increase from the current 7 billion to over 9 billion by the year 2043, thus making the problem more acute with the passing of time. Still another dilemma that compounds the quantity problem is the worsening of the quality of water as we continue to use it. Every time we use water, we add things to it, such as soaps, body fluids, shampoo, antibiotics, excreted prescriptions and medicines, and industrial and other pollutants. Many of these have potentially toxic components, depending on their concentrations. We know nothing about the toxicity of many of these compounds, and they accumulate in our water over time.

So it is not difficult to see that we face quantity as well as quality dilemmas. Fortunately, some solutions already exist for both of these problems, and new and more cost-effective solutions are being developed. So there is a glimmer of hope that humankind will be able to mitigate, if not solve, these problems. Efforts need to be made to implement solutions that we know work while we continue to develop new solutions.

Water Cycle—From Nature to Us and Back

Figure 2.2 shows a complete water cycle from nature to us and back from us to nature. With heat of the sun, water evaporates from oceans, rivers, and lakes to form clouds, which results in rain or snow under proper conditions of temperature and humidity. The water from rain and snowmelt either percolates into the ground or runs off into rivers, lakes, and streams. A portion of the water that percolates into the ground stays in the soil near the surface and either becomes a source of water for trees and plants or evaporates; a portion percolates deeper and becomes groundwater. The water the plants and trees use and the water evaporated from the soil surface is termed evapotranspiration.

The water that reaches the groundwater or lakes, rivers, and streams can be pumped out and treated to remove pollutants that it picks up on its journey through atmosphere, ground, or surface of the Earth. From there, the treated water is transported through a network of pipes to homes, industries, schools, hospitals, parks, and playgrounds. Some of this water percolates back into the ground—say, from landscaped areas, parks, and playgrounds—and some becomes wastewater after use in homes, industries, schools, hospitals, and other facilities. Now begins another journey for this wastewater, starting from the premises of users and going through yet another network of pipes, sewers, and pumping stations to a remote location, where wastewater treatment facilities are located.

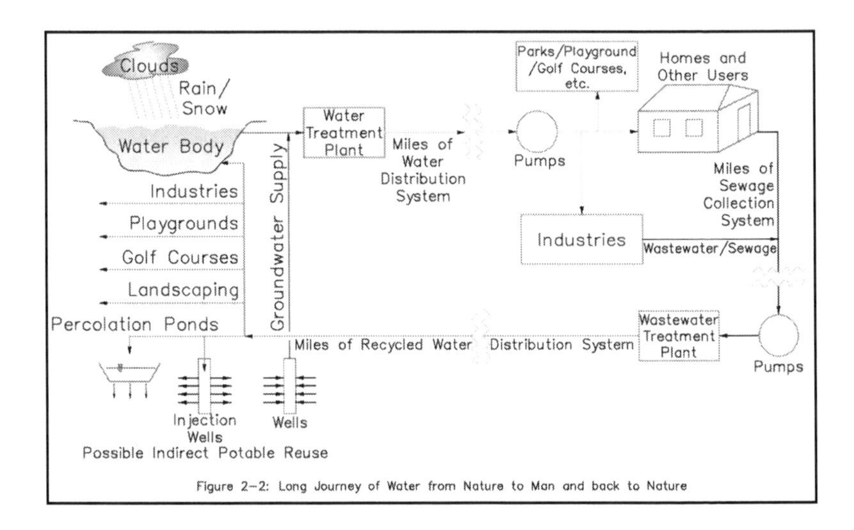

Figure 2-2: Long Journey of Water from Nature to Man and back to Nature

After treatment, this water—called wastewater effluent, recycled water, or reclaimed water—is reused or discharged back into a nearby stream, river, or ocean, or it is returned back into the ground via percolation ponds or deep injection wells similar in construction to water supply wells. The percolated and/or injected water can later be pumped out by another set of wells for indirect potable use. There are strict regulations for indirect potable use, including a mandatory separation distance between the injection wells and/or percolation ponds and the water supply wells; the time of travel for the recycled water between those two locations; and the type of soils through which recycled water must travel. These are all aimed at further ensuring the safety of this water for reuse.

What is not shown in the diagram is the accumulation of salts and minerals. Water naturally has minerals in it as a result of its passage over the land and through the soil and rocks; water dissolves minerals from the rocks. We add minerals to the water when we use it. We add soaps in our sinks and dishwashers, solids and mineralized liquids in our toilets, and fertilizers on our lawns, which can run off to rivers when we water our lawns. In cold northern climates, salt is frequently applied to streets and highways to control ice. When the ice melts, this too runs off to rivers and streams. Some homes have water softeners that use salt, which is flushed into the sewer when the softeners are regenerated. All of these minerals eventually find their way into rivers, streams, and lakes. When the water evaporates from lakes and streams, these minerals are left behind and, over time, make our water more mineralized and less desirable.

When we water our lawns or irrigate our crops, we use water containing minerals. A large portion of this water evaporates, leaving the salts and minerals behind. (Have you ever observed the ring of white deposits in a pan or teakettle? These are the minerals that were in the water that evaporated.) When rains come or we apply more water for irrigation, this water redissolves the minerals and transports them down into the ground. Eventually, the minerals reach the groundwater. When we pump the water up again to irrigate our crops, a large portion of those salts and minerals are left behind.

This vicious circle of salts continues until the water becomes unusable. There are many areas of the United States and the world that have groundwater so salty and so full of pollutants that it can no longer be used for drinking or irrigation.

Mineralization and water-quality deterioration are issues each of us must deal with. Eventually we will have to spend money and energy to remove these minerals. Each of us should be mindful of the water cycle and the salt cycle.

Thus the long arduous journey of water is now complete. In some cases, this can be hundreds of miles in rivers, streams, and pipes. Even though the water has received treatment at various stages of its long journey—while percolating into ground, at a water treatment plant as well as at a wastewater treatment plant—it is not as pure as it was when its journey started in the clouds. No treatment or combination of treatments accomplishes 100 percent removal of pollutants. Over time, the minerals and pollutants accumulate. Therefore, with each use, the quality of water deteriorates. With science, technology, and human ingenuity, this deterioration can be minimized so that it is of little or no concern.

Chapter 3

Wastewater Systems

Wastewater collection systems are either "separate" or "combined." Combined systems collect rainwater and street runoff along with sanitary wastewater from industries, commercial establishments, schools, hospitals, and residences. These are quite common in older cities in the United States; over 750 US cities have combined systems. These systems serve over 40 million people, and most of them are in the northeast, the Great Lakes region, and the Pacific Northwest (EPA, 2008). In separate systems, storm water and sanitary sewage are conveyed in separate piping systems—that is, a storm drain or storm sewer and a sanitary sewer or just a "sewer." Sewer pipes range in size from four inches in diameter to over ten feet in diameter, depending on the size of the population served.

The main sewers are usually in public streets; generally they start out small in size, usually eight inches in diameter, and flow by gravity. The pipes are sloped (slanted) to carry the collected wastewater with sufficient speed to transport the solids in it. Normally the pipes are only partly full to allow for ventilation. As more and more houses are connected, the pipes need to be larger and eventually dump into trunk or interceptor sewers. These larger sewers lead to a treatment plant. Manholes or access chambers are constructed at locations along the sewer pipe to allow for inspection and maintenance. Manholes are also located where main sewers connect to each other.

Where the ground is flat, sewer pipes can get very deep due to the need to maintain the proper slope; this makes construction difficult and expensive. Pumping or lift stations are needed to lift the sewage to a shallower sewer, where it can then begin its downhill travel again. Sometimes sewers are so deep they must be constructed as tunnels.

Individual houses and businesses are connected to a main sewer in the street by a small pipe called a house lateral or house connection. Typically, property owners maintain these connections, and cities are responsible only for the portion that is in the street.

In combined systems, storm-water pipes and roof drains are also connected to the sewer system.

Combined Sewer Systems

Combined sewer systems function well when carrying only sanitary wastewater. The sanitary wastewater flows to treatment facilities where it is treated and recycled or discharged. The problem arises when the pipes must handle rainfall runoff from streets, drainage from roofs and parking lots, snowmelt, etc. Even small to moderate rainfall events can overload the capacity of the pipes and overwhelm treatment facilities to the point that they are not able to process the water effectively.

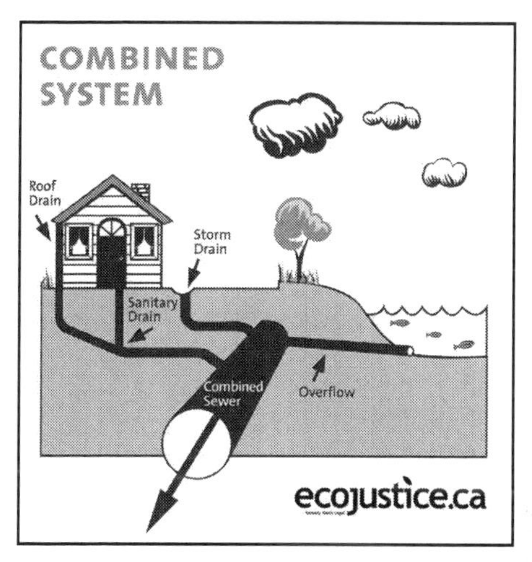

Schematic of a Combined Sewer System (Ecojustice)

Combined sewer systems were developed with overflow points to release extra water that cannot be carried or processed into creeks and rivers. Overflows from combined sewer systems are called combined sewer overflows (CSOs).

In the past, the water released into streams and rivers was a combination of sanitary sewage and storm water containing bacteria and viruses, oil and grease, fecal matter, industrial toxics, floatables, nutrients, etc. Although sanitary wastewater was diluted with storm water, it was still laden with bacteria and material that looked like sewage. Over time, cities installed screening facilities to remove the most objectionable material and, in some cases, installed makeshift disinfection processes. As cities grew, these releases or overflows became more frequent and more objectionable, resulting in numerous beach closures, fishing restrictions, and other water-quality impacts.

Communities with combined sewer systems are working diligently to correct the problems and minimize the CSOs. Sewer systems are being separated as much as possible; additional treatment plant capacity is being installed where cost-effective; storage is being developed to store these short-duration, high flows temporarily until there is capacity in the treatment systems to treat them after the storm has passed. A number of the larger cities, such as Chicago, Milwaukee, and Portland are constructing large caverns (tunnels) seven feet to over thirty-three feet in diameter and deep underground, which can be used to store CSO water. Structures are constructed to divert water into these tunnels whenever high flows occur. The water is stored there until the storm subsides, and the stored water is then pumped out to the treatment plant and treated.

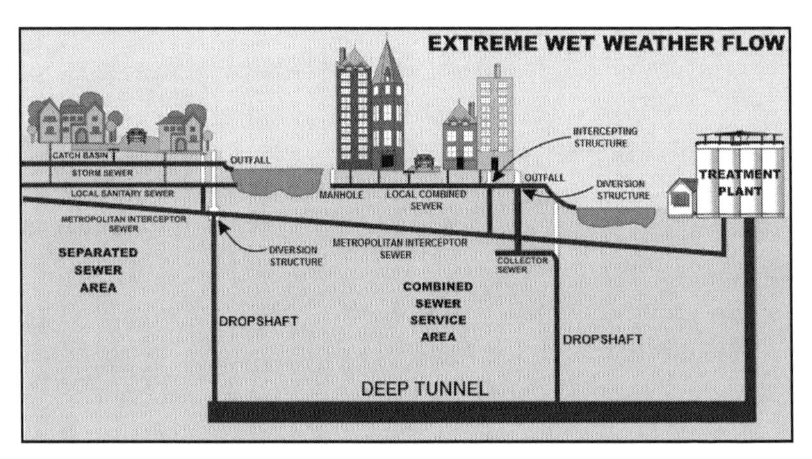

Deep Tunnel Project Schematic (Courtesy of Milwaukee Metropolitan Sewerage District)

These control strategies are working. Milwaukee's Metropolitan Sewerage District reports that before the tunnels were constructed, CSOs occurred fifty or sixty times each year. Over the nineteen-year period since the tunnels were added, this has been reduced to an average of 2.4 per year (Behm, 2013). This project cost $2.2 billion (Mueller, 2012).

Separate Sewer Systems

Separate sewers are designed to accommodate only sanitary wastewater. The piping systems are usually smaller than combined sewers, because they do not have to carry storm flows. These sewers lead directly to wastewater treatment plants and do not contain any overflow points.

Although they are supposed to carry only sanitary wastewater, the pipes get older, form cracks, and often leak at their joints. Tree roots grow through the cracks as they try to find a steady source of water. The roots grow, and the cracks get bigger. In addition to the roots causing blockages, the cracks allow shallow groundwater, called infiltration, to enter the pipes. This takes up capacity that could otherwise be used to carry sewage, and it increases a community's treatment costs.

When it rains, storm runoff water can enter a system through openings in sewer manhole covers at flooded intersections. In addition, there are often illegal yard and roof drains connected into systems. So when it rains, these separate sanitary sewer systems can experience relatively large increases in flows. These latter sources are called inflow. Often the term *inflow and infiltration,* or I/I, is used to describe this water. If excessive, it can overwhelm a sewer system and cause sanitary sewer overflows (SSOs). These are also the focus of considerable regulatory attention.

Overflowing Manhole

Dry-weather overflows can also occur as a result of broken or grease-plugged or root-plugged pipes, power outages preventing pumps from starting, etc. Cities are being fined for these occurrences and are now proactively cleaning and inspecting their sewer systems, establishing ordinances for restaurants to install grease interceptors, and providing emergency generators on pump stations.

Controlling I/I

I/I is expensive and burdensome. It takes up valuable space in a sewer system and costs money and energy to treat. Capacity has to be constructed at the treatment plant for this I/I. If I/I can be reduced, capacity is made available to treat sewage instead of rainwater and groundwater. Controlling I/I begins with locating and identifying the problem. Typically I/I is most severe in areas where pipes are older and are below the groundwater table. House connections are also a major source of I/I. Illegal connections can be identified by smoke testing a sewer system. Large fans blow colored smoke into a manhole, and unlawful connections or severely cracked pipe are found by watching the smoke rise. These can then be dug up and fixed.

Smoke Test (MAWSS)

Flow monitoring, using temporary flow meters, is useful in identifying areas where I/I is a problem. If it is a problem, the pipe can be replaced or rehabilitated. Rehabilitation can be accomplished by inserting a liner in the pipe or slipping another slightly smaller pipe inside the old pipe (called slip lining).

Cured-in-Place Slip Lining
(Photo by Dr. Chris Wood, PCWASA, 2013)

Asset Management

A wastewater collection and treatment system is a very valuable asset to a community and must be carefully managed. One of the first elements of asset management is to know what you have, how old is it, what condition is it in, and how long will it last. Once this is known, a budget can be developed to restore the asset and extend its useful life. Often the first step is to develop a Geographical Information System (GIS) map and database of the entire system. The database can contain information about pipes or other pieces of equipment, when they were installed, when they were last maintained, incidents of blockage, measured flows, etc.

Many agencies are implementing GIS to reduce the time it takes to identify a problem and take corrective action. Although developing a GIS is expensive, it is well worth the expenditure in the long run. A GIS is typically available to all staff on their smartphones or tablets, so they can use it anywhere they have access to the Internet. A system can include links to inventory, construction drawings, supplier and vendor information, etc. Those working in the field have instant access to this information, which saves valuable time, particularly in emergencies.

Condition assessment is typically performed using closed-circuit television cameras pulled through sewer pipes. Issues such as cracked pipes, illegal connections, sagging and clogged pipes, and pipes subject to frequent grease clogging or root growth can be easily identified. These videos can be linked into the GIS database along with the construction drawings for the system.

References

Behm, D. (2013, January 6). MMSD's Deep Tunnels Prevented Nearly 50 Sewer Overflows to Wateways. *Milwaukee Wisconsin Journal Sentinel*. Milwaukee, WI, USA: Milwaukee Journal Sentinel.

Ecojustice. (undated.). *Media Backgrounder Green Cities Great Lakes*. Retrieved November 3, 2013, from Ecojustice: http://www.ecojustice.ca/media-centre/media-backgrounder/media-backgrounder-the-green-infrastructure-report

EPA, U. (2008, October 15). *Combined Sewer Overflows*. Retrieved November 3, 2013, from U. S. EPA National Pollutant Discharge Elimination System: http://cfpub.epa.gov/npdes/home.cfm?program_id=5

MAWSS. (undated). *Private Sewer Lateral Replacement Program*. Retrieved May 15, 2014, from Mobile Area Water and Sewer System: http://www.mawss.com/private_sewer_later_replacement_program.html

MMSD. (undated). Milwaukee, WI.

MMSD. (undated.). *Overflows*. Retrieved November 3, 2013, from Milwaukee Metropolitan Sewerage District: http://www.mmsd.com/Overflow.aspx

Mueller, J. (2012). *An Evaluation Milwaukee Metropolitan Sewerage District*. Madison: WI State Auditor.

PCWASA. (2013, Jan 14). *PCWASA UTILIZES ENVIRONMENTALLY FRIENDLY INNOVATION TO REHAB SEWER LINES*. Retrieved May 15, 2014, from Peachtree City Water and Sewerage Authority: http://pcwasa.org/news/PCWASA-utilizes-environmentally-friendly-innovation-to-rehab-sewer-lines-

Raleigh NC, C. o. (undated.). *Sanitary Sewer Overflows, What You Should Know.* Retrieved November 3, 2013, from City of Raleigh: http://www.raleighnc.gov/environment/content/PubUtilAdmin/Articles/SanitarySewerOverflows.html

Chapter 4

Wastewater Characteristics and Flow

Introduction

This chapter discusses wastewater characteristics in terms of commonly used parameters as well as the quantity of wastewater generated by a community. Both the characteristics and the flow are important parameters in the design of municipal and industrial wastewater treatment facilities.

Wastewater Characteristics

Wastewater is about 99.95 percent water by weight. The rest, 0.05 percent, is material dissolved or suspended in the wastewater. A generally accepted estimate is that each individual, on a national (United States) average, generates approximately eighty to one hundred gallons of wastewater per day. A typical household in the United States generates about 250 to 350 gallons of wastewater per day. The pollutants of concern within the wastewater consist of both organic and inorganic materials. The organics in wastewater originate from our bodily excretions (fecal matter) as well as ground garbage from kitchen sinks. Industries such as breweries, food processing plants, and dairies generate their own unique wastewater organics, which are characteristic of the raw materials they use or the products they produce.

Organic compounds are largely a combination of carbon, hydrogen, oxygen, and usually nitrogen. Other important elements such as sulfur, phosphorus, and iron may also be present in trace amounts. The principal groups of organic substances found in wastewater are proteins (40 to 60 percent), carbohydrates (25 to 50 percent), and fats and oils (10 percent). The use of water in a municipality also adds inorganic compounds, such as ammonia, sulfates, chlorides, phosphorus, and toxic substances, such as cadmium, chromium, and zinc.

Some organics and inorganics are present in the wastewater as suspended matter; the rest are in solution. Most of the suspended solids can be removed simply by allowing the wastewater to stand undisturbed for a period of time to permit the solids to settle. The dissolved organic and inorganic pollutants, however, are more difficult to remove.

Of the organics found in wastewater, a substantial portion consists of biodegradable materials—those that serve as food sources for bacteria and other microorganisms. The biological breakdown of these materials by microorganisms consumes oxygen.

The amount of oxygen required to stabilize biodegradable organics is measured by the biochemical oxygen demand

(BOD) test. The BOD is an indirect measure of the organic strength of the wastewater. It is founded on the principle that the greater the amount of biodegradable organic matter in the wastewater, the higher the BOD—that is, more oxygen is taken from the water by the microorganisms to break down (oxidize) the organics. This parameter is the most widely used to measure the organic content of the wastewater. It is used in sizing treatment facilities and in predicting the effects of treated wastewater discharges on receiving waters such as streams and rivers. If the oxygen demand remaining in the treated wastewater exceeds the oxygen replenishment rate from the atmosphere to the lake, river, or stream, the oxygen in the water may be completely depleted, and the stream or lake will become septic at or downstream from the wastewater discharge point. A septic stream is easily identified by its murky black color and noticeable odor. Ultimately, due to re-aeration from the atmosphere, particularly in turbulent streams, the stream returns to a "clean" state unless it receives another slug of pollution from another source. (See figure 4.1.)

Some of the organics in wastewaters are not readily biologically degradable and thus are not part of the BOD. These nonbiodegradable organics pass through a treatment system generally in the same concentration they entered it. Some of these non-degradable organics, such as many pesticides, can have adverse long-term effects and can contribute to taste, odor, and other problems in receiving waters to which wastewater—treated and untreated—is discharged. The chemical oxygen demand (COD) test is used to measure the biodegradable and nonbiodegradable quantities of these materials in water. Since the COD value also includes biologically degradable materials, the COD concentration is always higher than BOD—typically about twice the BOD for municipal wastewater.

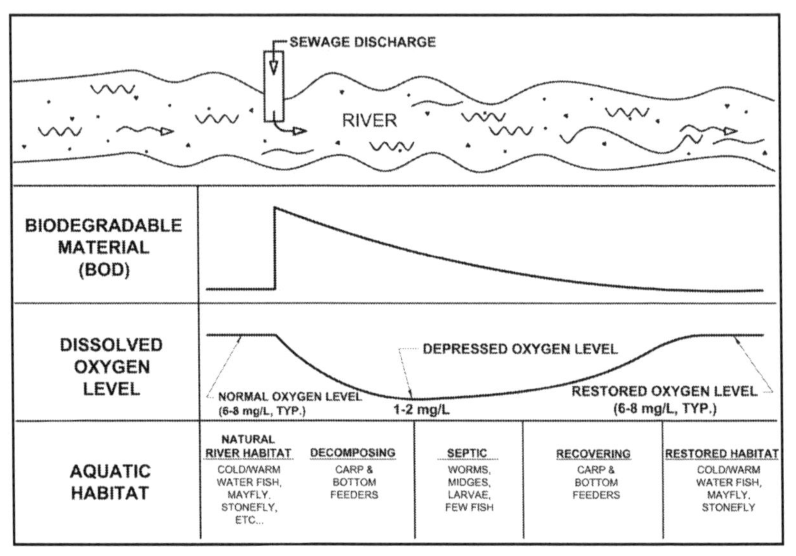

Figure 4.1 Impact of Sewage Discharge on a Receiving Stream

Some COD-causing materials are organics that are very resistant to breakdown in the environment. They are of particular concern where water is used for municipal water supplies downstream. Fortunately, treatment techniques are available for removing wastewater COD as well as BOD.

Municipal wastewater also contains microorganism, including bacteria and viruses, that can transmit diseases. (Organisms that cause disease in humans are called pathogens.) This can be especially critical if the receiving water is used for recreation or as a drinking-water supply. These pathogens must be inactivated or "killed off" before the wastewater is discharged.

Wastewater also contains two other important elements: phosphorus and nitrogen. The phosphorus is primarily in the form of phosphates, which originate from our bodily wastes; it is in the food we eat. It used to be in high concentrations in detergents but this is no longer the case, now that phosphorus-free detergents are standard.

Nitrogen in raw wastewater is in the form of ammonia and organic nitrogen. Organic nitrogen is from urea in the urine and organic compounds in fecal matter. Ammonia results from the breakdown of urea (ever smell a saturated baby diaper on a nice warm day?). It can exist in water in two forms: dissolved ammonia gas and the ammonium ion. The amount of each depends on the pH of the water. Dissolved ammonia gas (called unionized ammonia) is very toxic to fish and aquatic animals, and as a result is regulated to very low concentrations. The ammonium ion (called ionized ammonia) is not toxic but has some problematic reactions in water. It can be oxidized to nitrate by microorganisms in water. In the process of doing this (called nitrification), large amounts of oxygen are consumed by the microorganisms—similar to that described above for BOD, only much more so. This process can rapidly deplete the oxygen in a river or lake and cause fish kills. Many treatment facilities are mandated to install additional processes or make process modifications to remove ammonia. This can be quite costly from a capital standpoint and also from an energy and operational cost standpoint.

Nitrogen and phosphorus are also called nutrients, since they can stimulate the undesirable growths of algae in lakes and streams. These algal growths can also cause unpleasant tastes and odors in water supplies and operating problems in downstream water-treatment plants. Algae are photosynthetic organisms. During the daylight hours, they get energy from sunlight and produce oxygen far in excess of what they need to live. This excess oxygen dissolves in the water, which is good. But at night, algae need oxygen to continue living, so they take oxygen from the water. This is not good, as it can deplete the oxygen needed by fish and other higher life forms. Fish kills can result. Algae also exert a significant oxygen demand when they die and settle to the bottom of a river or lake. Where receiving waters are particularly prone to algal growth because of the nutrient composition of the water in the river or lake, the removal of phosphorus and possibly nitrogen may be desirable.

The toxic nature of heavy metals, such as cadmium, chromium, zinc, and lead, can interfere with biological wastewater treatment processes. If these metals enter the receiving water in sufficiently high concentrations, they can cause fish kills and can create a problem in downstream water supplies. In smaller quantities, they may not cause immediate fish kills, but can enter the aquatic food chain, where they can accumulate and cause long-term health and environmental problems.

Pharmaceuticals and personal care products (PPCPs) are contained in municipal wastewater in very low concentrations, but they are particularly alarming since many of these are endocrine disruptors. These compounds (along with a number of others, including some pesticides) mimic normal hormonal behavior and turn on, turn off, or modify signals that hormones carry. As a result, many of these substances have been linked to developmental, reproductive, and other changes in wildlife and laboratory animals (NIEHS, 2010).

Our waste products, feces, and urine are the primary source of these compounds. And many agencies have a campaign to inform their customers not to dump prescription drugs "down the drain."

Typical concentrations of pollutants in raw, untreated municipal wastewaters are as follows:

- BOD: 150 to 250 mg/L
- COD: 300 to 400 mg/L
- Suspended solids: 150 to 250 mg/L
- Phosphorus: 5 to 10 mg/L
- Nitrogen: 25 to 40 mg/L
- Total Dissolved Solids (TDS): 150 to 300 mg/L more than the TDS in the water supply

These concentrations vary from one community to another and must be established by analyzing samples of wastewater for design of any treatment facility.

Wastewater Flow Rates

As pointed out above, each individual (on a national average in the United States) contributes approximately eighty to one hundred gallons per day of wastewater to a sewer system. Therefore, a city of about ten thousand people will contribute about 1 million gallons per day of wastewater. In addition, allowance must be made for the industrial contribution to the sewer system, although it may be somewhat difficult to estimate. Depending on the type of industry, the contribution can vary significantly in quality and quantity. As a general rule, the industrial contribution to municipal sewer systems must be controlled and monitored. (It is often easier to pre-treat industrial wastes at their source than to allow them to enter the municipal sewer system.)

However, the contribution to the sewer system is not constant during the day, nor is it constant during the entire year. There are both diurnal (twice per day peak) and seasonal variations in wastewater generation in each community, depending on its size, its nature, its social characteristics, and the length of its sewer system. (See figure 4.2.) In addition, depending on the age of the sewers and their condition, there are some extraneous flows that can enter the system. These are groundwater if sewers are below the groundwater table and the sewers have leaky joints (called infiltration) or could be from unauthorized connections from roofs and yards (called inflow).

Together, these are known as infiltration and inflow (I/I), as discussed in chapter 3, and can be significant contributors to wastewater flow. Communities are constantly taking steps to remedy this problem by replacing old, leaky sewers as well as policing and disallowing unauthorized connections. If not corrected—the communities realize this—the wastewater treatment plants can get overwhelmed with increased flow, which would affect the quality of treatment provided at the plant. In addition, sewers are often overloaded with excessive flow, with sewage spilling out of the manholes and causing unhealthful and nuisance conditions.

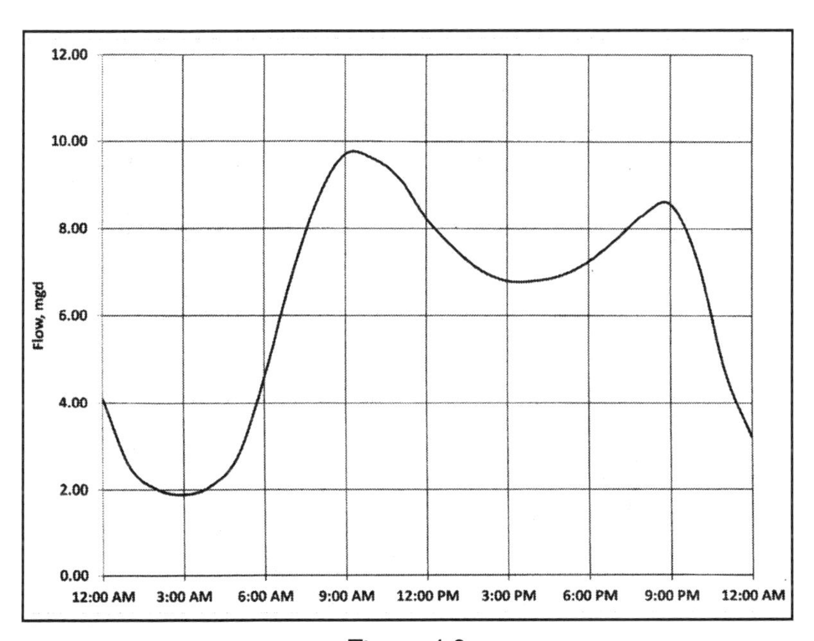

Figure 4.2
Typical Variation in Municipal Wastewater
Flow Rate During the Day

Even without I/I, flow variations occur in all sewer systems and at treatment plants. Generally, minimum flows occur during the early morning hours when water consumption is the lowest. The first peak flow occurs in the late morning hours when wastewater from the morning water use reaches the treatment plant. A second peak generally occurs in the early evening hours—say, from seven to nine, when water use reaches another high. It is easy to visualize that this will vary between communities depending on their size and the length of their sewer systems, since wastewater does not reach a treatment plant instantaneously after water use by members of the community. Figure 4.2 shows a typical diurnal variation pattern.

In addition to diurnal variations, there are weekly and seasonal variations as well. For example, most households do their laundry over the weekend. Similarly, some communities are resort towns visited by tourists during certain times of the year and are less populated at other times of the year. For example, tourists frequent Palm Springs, California, during winter while the population is much less during hot summers.

A treatment plant must be able to handle these flow variations in its design without compromising the level of treatment required. So, not only is the average flow at the treatment plant important for a good design, but equally important are the flow variations, both daily and seasonal, that are expected at the plant.

References

NIEHS. (2010, May). *Endocrine Disruptors*. Retrieved May 15, 2014, from National Institutue of Environmental Health Sciences: http://www.niehs.nih.gov/health/materials/endocrine_disruptors_508.pdf

Chapter 5

Wastewater Treatment and Disposal—Early Days to the Present and Beyond

Introduction

The desire to stay healthy and happy and to live in a clean environment is as old as humanity. We may not have known how to achieve it through the ages, but we have always had that desire. One aspect of this is the knowledge of how to treat properly and dispose of the wastes—liquid, solids, and others—generated by our activities and just living.

This chapter primarily deals with the wastewater generated on a day-to-day basis and how it was handled through the ages. One of the earliest systems of wastewater management was constructed in Mohenjo-Daro near the river Indus (Pakistan) in about 1500 BC. Some centuries later, the river changed its course, and the town was abandoned. Over time, it became covered by sand and then was rediscovered in the 1930s. Public buildings and private houses were equipped with toilets. Both rainwater and water used for washing and bathing flowed through special conduits into canals with a slope so that it could flow into the river by gravity (Wiesmann et al., 2007).

Another important evidence of wastewater management is seen in Rome. A ditch was used from 500 BC as a collector for wastewater. Due to population growth, it was enlarged in the following centuries, extended, and roofed over, with wastewater flowing by gravity to the river Tiber near the Ponto Palatino. Later, in 31 BC to AD 193, the canal could be traveled by boat and could be entered through manholes (Unknown, Fundamentals of Biological Wastewater Treatment, 2007). Roman waste management practices were the most developed of any civilization before the nineteenth century. The Romans were very advanced technologically. We can still see their buildings, roads, and aqueducts today. The aqueducts were used to transport water to the cities. The water was used for baths, fountains, and public toilets, and for flushing sewers. Through flushing, the wastes were transported to the nearby river Tiber via open sewers as early as the sixth century BC, as reported in "The Evolution of Sewage Treatment," an unpublished paper by an unknown author (which also provides the information in the next paragraph).

As societies moved from nomadic cultures to building more permanent sites, the concern over waste disposal (solid and liquid) became an important one. This issue was dealt with in different ways over time as knowledge was gained, lost, and regained. The primary concern was to address the issue of infectious diseases, not necessarily chronic health risks or having a healthy environment. In the Egyptian city of Herakopolis, the average person threw his or her waste into the streets, not knowing any better. However, in the elite and religious quarters, a deliberate effort was made to remove all wastes to locations outside the living areas, which usually meant the rivers and other water bodies. Mosaic law (the Law of Moses, 1300 BC) required the Israelites "to remove your own waste and bury it in the earth." The Minoan culture on the island of Crete (Greece) between 1500 and 1700 BC had a highly developed waste management system. The capital city, Knossos, had a central courtyard with baths that were filled and emptied through terra-cotta pipes. They also had community flushing toilets with wooden seats and an overhead water reservoir.

In the Middle Ages, AD 500 to 1500, the Roman Empire fell, turning the urban society into a rural one. By 500, sanitation technology had entered its dark ages. And it remained there for almost a thousand years. There was a massive depopulation of Rome and most of the western empire. The reduced population density rendered traditional methods of waste disposal—tossing it into the street—more viable. The demise in sanitation brought back outhouses, open trenches, and chamber pots at all levels of society.

These historical examples are only to illustrate the development of wastewater management through early times and the innate human desire to live in a clean environment as best as we can with the knowledge and resources at our disposal. Time moved on, advancing the knowledge, the resources, and the desire to improve living conditions, health, and well-being.

The Birth of Microbiology

Antoni van Leeuwenhock (1632–1723), a cloth merchant in Delft, a city in the Netherlands, was the first person to grind simple lenses and build the first microscope. He constructed many of these microscopes himself. The best of them had a magnification of 200x. Using his home-built microscopes, he observed small things in water droplets. In 1674, he reported "small animals" and small globules in water drops. Later, in 1683, he prepared a slide of these small animals from the scrapings of his own teeth. He reported to the Royal Society of London seeing three shapes of these animals, which were labeled as cocci (spheres), rods, and helical (spiral or spirilla). This is what we know even today as the three shapes of microorganisms (Byrdon & Williams, 1969).

This knowledge was expanded by Benjamin Marten in 1720 (London), when he theorized, based on a visual examination of little animals or little beasties in the excrement of his patients, that the cause of tuberculosis (called consumption of the lungs) was these animalcules/microorganisms, which were believed to travel to the lungs with the circulation of blood (Wiesmann et al., 2007)

This knowledge expanded over the years through the work of scientists, public health officials, physicians, microbiologists, and sanitary engineers. These efforts led to the conclusion that poor sanitation, untreated sewage, and trash flowing into rivers and other water bodies, serving as the source of drinking water, were causing diseases, particularly cholera, typhoid, infectious hepatitis and gastroenteritis, if the water was not treated. Cholera epidemics in London (1848 and

1854) caused by drinking water taken from sewage-contaminated parts of River Thames and from the Broad Street Public Well, and numerous outbreaks of diseases, including cholera in 1832, 1849, and 1866 as well as typhoid in 1848 in many US cities, are attributed to a variety of reasons, including unsanitary conditions and punishment from God (Unknown, *Urban Wastewater Management in the United States: Past, Present and Future*).

The experience gained from these frequent epidemics improved the understanding of cholera and other waterborne diseases and their corresponding etiology. It was concluded by rational-minded scientists and public health officials that there was a strong correlation between these diseases and unsanitary conditions, including the waste discharges into rivers and other drinking-water sources. No longer could it be assumed that a river could be loaded with untreated wastewater and cleaned with the self-purification properties of the receiving waters (Wiesmann et al., 2007)

Some Important Developments

As the population grew with increased urbanization and the concentration of people in major population centers, the problem of wastewater disposal multiplied. Human waste could no longer be disposed of in the old-fashioned way—that is, hauling it away to remote, outside-the-city locations. By the mid-nineteenth century, engineers, public health officials, and the public were searching for alternative ways to dispose of wastewater. The re-advent of a water carriage system similar to the ones used by Romans and the creation of the modern toilet/water closet in Europe and the United States were two major developments in achieving that objective. Now human waste products could be diluted with large quantities of water and carried through sewers with adequate velocity to transport solids to a centralized location outside the population centers, where it could be disposed of in a river or a suitable water body.

But discharging untreated wastewater into rivers polluted the receiving water, causing aesthetic problems and odors at the point of discharge, fish kills, and a loss of the recreational values of those waters. Using these waters for drinking, with little or no treatment, caused serious health problems that needed to be tackled with science, financial resources, and political will. How to treat that wastewater before discharge continued to be an unsolved question until the early to mid-twentieth century.

As these developments were occurring, in 1913 and 1914, Edward Arden and William T. Lockett, from the River Committee of the Manchester Corporation in the United Kingdom, were in their laboratory, proving experimentally the role of bacteria in the treatment of sewage. A higher concentration of bacteria reduced the time for treatment in their experiments from, say, twenty-four hours to eight hours. This was the "birth" of wastewater treatment as we know it today.

The upshot of their research is that if a proper environment is created—pH, salinity, temperature, concentration of microorganisms/bacteria, oxygen levels, nutrients, etc.—in reactors/tanks at a wastewater treatment plant, wastewater can be treated. Fortunately, all wastewaters from domestic sources, such as homes and schools, have the right salinity, nutrients, temperature, and pH. The only parameters that need to be controlled are oxygen level and concentration of microorganisms; the latter are ubiquitous and most certainly in wastewaters.

All biological processes in use today for wastewater treatment are essentially based on this simple principle. Many treatment processes have been developed based on it and will continue to be improved, developed, and invented, but the principle is the same: create a suitable environment as defined and measured by the necessary parameters, and the microorganisms will take it from there to provide most of the treatment.

This is, of course, a simplified picture of the treatment of wastewaters. Wastewaters today—particularly generated by industries, chemical plants, and drug manufacturers—are complex in nature and chemical composition. And regulators are demanding higher levels of treatment to achieve removals of these micropollutants because of their known and unknown effects on human health and the environment. So biological treatment, which remains the workhorse of all treatment processes, is followed by other treatment processes to accomplish these removals. These may include membrane processes where the biologically treated wastewater is passed through porous media or membranes with small pores to remove or filter out these micropollutants.

Several types of media and membranes—proprietary and nonproprietary—are marketed under different process names such as filtration, ultrafiltration, microfiltration, nanofiltration, and reverse osmosis, depending on the size of particles, including bacteria, viruses, protozoa, and others that are the targets for removal. Essentially wastewater can be treated to any level desired by a combination of these processes, depending on the intended reuse of the water.

Some Early Philosophies on Treatment

In spite of a large body of evidence connecting the discharge of untreated wastewater into rivers with diseases, the discussion now was focused on whether both wastewater and drinking water should be treated or only the drinking water. While boards of health in the United States were generally in favor of treating both drinking water and wastewater, consulting engineers and municipalities were in favor of only drinking-water treatment for economic reasons. A noted engineer, Allen Hazen, expressed this opinion in 1903 in an issue of the *Engineering News Record*: "It is often more equitable to all concerned for an upper riparian city to discharge its sewage into a stream and a lower riparian city to filter the water of the same stream for a domestic supply, than for the former city to be forced to put in wastewater treatment works" (Burian et al., 2000).

This argument was quite successful in the early twentieth century. By 1905, more than 95 percent of the urban population discharged its wastewater untreated to waterways. In 1924, more than 88 percent of the population in cities of over 100,000 continued to discharge their wastewater into the waterways (Unknown, *Urban Wastewater Management in the United States: Past, Present and Future*).

Hazen promoted the mind-set that the treatment of the drinking water is all that is needed, and it lasted for decades. In the mid-1920s, H. W. Streeter, a sanitary engineer, and E. B. Phelps, a consultant to the US Public Health Service, published a seminal work on the biochemical oxidation of organic matter discharged into the Ohio River and the re-aeration and recovery of the river over time (Streeter, 1925). The Streeter-Phelps equations, as we call them today, clearly demonstrated a limit to the amount of organic matter that can be discharged into a river before dissolved oxygen levels necessary for fish and aquatic life become depleted. This simple model is the basis of today's sophisticated, computer-based water-quality models.

An interesting, shocking, but true episode documenting the extent of pollution in our waterways is worth mentioning again. The Cuyahoga River in Cleveland, Ohio, caught fire on June 22, 1969, due to oil-soaked debris at the bottom of the river, undoubtedly discharged by industries along the river. Fish were found to have tumors, and the water had turned "dark and soupy."

The desire for water-quality preservation and resource conservation now began to gain a foothold. Many began to realize that both water treatment and wastewater treatment are important. The recreational and aesthetic quality of water and the protection of aquatic life were just as important as the health-related aspect of our waters.

As all these concerns were taking hold in our society, the US government took a big leap forward by passing the 1972 Water Pollution Control Amendment to the Clean Water Act.

The law set the goal of eliminating all water pollution by 1985 and authorized expenditures of $24.6 billion for research and grants for constructing wastewater collection and treatment facilities. Regulations were developed for industrial and agricultural discharges. The passage of this landmark legislation and subsequent amendments and funding were the beginning of advanced wastewater treatment in the United States (Burian et al., 2000).

Among other things, the Clean Water Act established the National Pollutant Discharge Elimination System (NPDES), which required each point source discharger to waters in the United States to obtain a discharge permit. (A "point source" is any pipe, conduit, or channel that discharges pollutants to a river or waterway.) The act required the establishment of limits on industrial wastewater that could be discharged into public sewer systems, mandated a minimum of secondary treatment for municipal treatment facilities, prohibited the dumping of solid waste and treatment-plant residuals into the ocean, and created a process for establishing treated wastewater discharge limits. The act also required each state to implement a continuous water-quality planning process for all waters in the state, which identified the uses of water, such as drinking water supplies, agricultural water supplies, fish and wildlife habitat, etc.—often called beneficial uses. Once the beneficial uses were established, water-quality standards, often called water-quality objectives, were set by states to protect those uses. Permit limits for discharges into water are set on a case-by-case basis, to ensure those water-quality standards or objectives are maintained (33 U.S.C., 2002).

In the event a stream or river still could not meet the standard to preserve beneficial use, the act required the establishment of total maximum daily loads (TMDLs) for the pollutant(s) of concern and the allocation of a limiting load to each of the current dischargers. Some "load" was reserved for future dischargers.

This process has continued since 1972. The planning process, beneficial uses, and water-quality standards are documented for major watersheds in each state's Basin Plans.

Other Related Developments

Now that wastewater treatment was the law of the land, the focus shifted to developing cost-effective technologies for treating wastewater. On-site treatment facilities employing septic tanks, seepage pits, and Imhoff tanks for larger yet individual institutions such as schools and hospitals were still being built, which did not necessarily solve the problem as the law probably intended. These facilities still needed to be cleaned frequently, yet were often ignored, causing spills, ponding, aesthetically unappealing conditions, and perhaps, in the worst case, disease.

This practice continues today in some suburban areas where the density of housing has made on-site systems a source of pollution. Even with pollution evident, the residents in these suburban areas are reluctant to give up their septic tanks. They are not interested in building major infrastructures (sewers and the treatment plants) and incurring their share of the large upfront costs required for these major projects. They also fear that these projects will attract more people to their localities, which will diminish the exclusive nature of their neighborhoods.

However, despite the concerns of the elite few, the Water Pollution Control Act Amendments of 1972 provided a significant legal and socioeconomic impetus to the planning, design, and construction of wastewater collection and treatment facilities across the nation. This was accompanied by a number of studies and a great deal of research in developing cost-effective technologies for wastewater treatment in centralized facilities. These included physical treatment processes that used large chemical dosages (lime and ferric chloride) to settle BOD and organic pollutants; granular activated carbon to remove soluble organics; ammonia stripping towers for removing ammonia at a pH of 10.5 to 11, thus requiring high dosages of lime; ion exchange resins to remove ammonia or nitrates; processes employing heavy dosages of chlorine to oxidize sludges and remove nitrogen; and several proprietary processes for treating both liquid waste as well as solid waste—that is, sludges.

Many of these processes, some of them funded at essentially 100 percent under the Innovative and Alternative Technologies Program—a 1977 amendment to the Clean Water Act—were abandoned due to being inefficient, costly, and difficult to operate or because newer technologies, subsequently developed, were more cost-effective. But these setbacks also had a positive effect. Newer processes that were cost-effective and easy to operate and control were used at numerous plants all across the United States—often with public funds and subsidies.

Most recently, the focus has shifted from the treatment of wastewater to its reuse, since planners and public health officials are realizing that water supplies are not unlimited and that population is increasing with the associated increased demand on water. They are also realizing that sources of water are getting more polluted due to both increased population and increasing industrial, commercial, and agricultural discharges to rivers and water bodies. The need to develop more sophisticated and complex technologies is also stemming from our ability now, with new equipment and analytical techniques, to detect pollutants at parts per trillion levels (ppt). So the need to reuse is preempting the need to treat to previous standards. This is further challenging our researchers and water-quality scientists to continue to develop new technologies for wastewater treatment, including reuse.

Concurrent with wastewater treatment is the growing need for the development of technologies for sludge/biosolids handling and treatment, which are byproducts of wastewater treatment. Earlier, biosolids could be landfilled or applied to land. That is now frowned upon by environmentalists and citizens, since landfills can be sources of odor, can pollute ground water through leaching, can attract rodents and flies, can degrade property values, and can be perceived as causing health problems to those living in their proximity. In addition, landfills are filling up to their capacity, and it is difficult, if not impossible, to be permitted by local authorities to create a new one.

EPA 503 Regulations (40 CFR 503 Regulations), adopted in 1987, tried to address these concerns by regulating the handling and disposal of sludges/biosolids. However, in several cases, local authorities have passed their own regulations, which are stricter than 503 Regulations and have thus challenged these federal regulations in the courts; doing so has delayed and even torpedoed the implementation of the 503 Regulations. This has given strong momentum to the development of new technologies for the handling, treatment, and disposal of sludges that are more environmentally friendly. Reusing sewage solids, or minimizing the quantities of the sewage residuals before disposal, is the main objective of these innovations and developments. Research in this arena will continue for years to come. In the meantime, technologies will be developed; some will fail, some will be modified, and some will be reinvented.

More recent focuses have been on green technologies and on designing and operating a sustainable wastewater treatment plant. Words such as *green* have become fashionable. Politicians, scientists, engineers, teachers, and social scientists use the words *eco-awareness* and *sustainability* in setting funding and research priorities and educating future generations. Policymakers in Washington, D.C., and state capitals are committing billions of dollars to researching and implementing projects in the next decade and beyond, for renewable energy sources that will save the Earth's resources, such as water, energy, and raw materials, preserving them for future generations (Arora, 2009). These include the resources used during equipment manufacturing and construction and operation of treatment plants. New technologies, therefore, will be judged and funded based on green criteria in addition to economics.

The Future of Wastewater Treatment

It appears that biological processes will continue to be the main ones for wastewater treatment in the foreseeable future. No doubt, there will be improvements to these processes as researchers and wastewater professionals discover new species of microorganisms and as our understanding of the kinetics of different biological reactions responsible for treatment improves and advances.

Similar advancements will take place in our handling and disposal or reuse of biosolids. The advances will be aimed at resource conservation, enhanced production of digester gas in anaerobic digestion, cogeneration using this biogas, and minimization of the quantities of residuals requiring off-site disposal.

Water reuse will increasingly become a necessity. With that as a primary objective, new and cost- effective technologies will be developed to remove micro-pollutants – antibiotics, hormones, pharmaceutical compounds, pesticides, and personal care products, that are being detected in our wastewaters and water supplies with their unknown and less known long term effects on our health. Therefore, all future improvements and developments will focus on technologies that will enhance treatment and make water reuse economically viable.

The future of wastewater treatment is bright—and also challenging.

References

33 U.S.C., 1. e. (2002, November 27). Federal Water Pollution Control Ac t[As Amended Through P.L. 107–303, November 27, 2002].

Arden, E., & Lockett, W. T. (1914). Experiments on the Oxidationof Sewage without the Aid of Filters. *Journal Soc. Chemical Ind., 33*, 523-539.

Arora, M. L. (2009, September). Greener Plants. *Water Environment & Technology* .

Burian, S. J., Nix, S. J., Pitt, R. E., & Durrans, S. R. (2000). Urban Wastewater Management in the United States: Past Present and Future. *Journal of Urban Technology, 7* (3), 33-62.

Byrdon, K. L., & Williams, R. P. (1969). *Microbiology.* London: MacMillan.

Foil, J. L., Cerwick, J. A., & White, J. E. (1993, December). Collection Systems Past and Present, A Historical Perspective of Design, Operation and Maintenance. *Operations Forum, 10* (No. 12).

Haworth, J. (1922). Bio-aeration at Sheffield. *Proc. Assoc. Mgr Sewage Disposal Works,* (pp. 83-88).

Streeter, H. W. (1925). A Study of the Pollution and Natural Purification of the Ohio River. *Public Health Bulletin, 146.*

Wiesmann .U, Choi I.S., Dombrowski E.-M. (2007). *Fundamentals of Biological Wastewater Treatment.* Wiley : VcH Verlang GmbH & Company, Weinheim, Germany.

Unknown. *The Evolution of Sewage Treatment.*

Water Environment Federation, WEF. (2000). *Aerobic Fixed-growth Reactors.*

Water Pollution Control Federation, WPCF. (1997). *Wastewater Treatment Plant Design, Manual of Practice No. 8.* Water Pollution Control Federation.

Other Works of Interest

Arden, E., and W. T. Lockett. 1914. Experiments on the Oxidation of Sewage without the aid of filters, *Journal Soc. Chemical. Ind.* 33, 523-539. Quoted in "Fundamentals of Biological Wastewater Treatment." Wiley—VCH Verlag GmbH & Co. KGaA, 2007.

Haworth, J. 1922. Bio-aeration at Sheffield. *Proc. Assoc. Mgr. Sewage Disposal Works*, 83-88.

"Wastewater Treatment Plant Design." 1997. WPCF Manual of Practice, no. 8.

Water Environment Federation. "Aerobic fixed—Growth Reactors." 2000.

Metcalf & Eddy, Inc. 1991. "Wastewater Engineering Treatment, Disposal and Reuse." 3rd ed. McGraw Hill.

Chapter 6

Effluent Limits

The federal Water Pollution Control Act Amendments of 1972, commonly called the Clean Water Act (CWA), set up the structure for limiting the discharge of pollutants into oceans, rivers, and streams. The act set a national goal to "restore and maintain the chemical, physical, and biological integrity of the Nation's waters" with interim goals that all waters be fishable and swimmable where possible (33 U.S.C. 2002). The act gave the US Environmental Protection Agency (EPA) authority to set standards and regulations for pollutants and to establish effluent limits, but the states would administer and enforce the CWA programs. The CWA established the National Pollutant Discharge Elimination System (NPDES) program, wherein all dischargers needed to have a valid permit for their discharge.

Point Source Pollution
Source: (NOAA, Non-Point Source Pollution, Categories of Pollution: Point Sources 2008)

Pollution sources are identified as "point source" and "non-point source." Point sources typically come from a single pipe or single identifiable source. Municipal wastewater treatment plants and industrial discharges are good examples of point source discharges. These sources are easily regulated and controlled. Nonpoint sources of pollution come from spread-out or diffuse sources, such as runoff from urban and agricultural areas. They are much more difficult to control.

Nonpoint Source Pollution
Source: (NOAA, Welcome to Non-Point Source Pollution, Non-point Source Pollution Agricultural Operations 2008)

Surface Water Discharges

The CWA essentially mandated that all wastewater treatment plant discharges provide at least secondary treatment and that industrial dischargers provide pretreatment to remove toxic pollutants from their effluent before discharging to either a water body or a public sewer system. There are some exceptions that allow for equivalent to secondary treatment. (USEPA, NPDES Permit Writer's Manual 2010). Table 6.1 presents the secondary treatment effluent requirements.

Table 6.1
USEPA Secondary Treatment Standards
(USEPA, NPDES Permit Writer's Manual 2010)

Constituent	30-day Average	7-day Average
Biochemical Oxygen Demand, BOD, mg/L	30	45
Total Suspended Solids, TSS, mg/L	30	45
pH	6-9	—
Minimum percent removal	85 percent of BOD and TSS	—

In addition to the limits in table 6.1, the USEPA can set limits on other parameters, including bacteria, depending on the receiving water body.

The process of setting the effluent quality or limits begins with the identification of the beneficial uses of the water body to be protected. There are a number of beneficial uses, including public water supply, swimming and recreation, cold-water fisheries, shellfish harvesting, and agricultural water supplies. Once the beneficial uses have been established, water-quality criteria or objectives or standards are set for the water body to protect and maintain those beneficial uses. Discharge limits for various constituents are set for each discharger so that the water-quality objectives are maintained. Where the existing water quality is very good, states have antidegradation policies that state that the existing water quality cannot be made worse as a result of a discharge. In these cases, the constituents in the discharge could be limited to very low concentrations.

There are many cases in the United States where rivers, streams, and other water bodies do not meet even the minimum requirements for the designated beneficial use. For these waters, the specific constituents that cause the impairment are identified, and the water bodies are placed on the 303d List, named for the section in the CWA legislation. This list must be reviewed and updated on a regular basis.

A study is conducted by the state to determine the total amount of the critical constituent that can be accommodated and still meet the water-quality objective needed to protect the beneficial use. This amount is called the Total Maximum Daily Load (TMDL). This is the maximum amount of any given constituent that a water body, such as a river or lake, can receive daily and still meet the established water-quality standards. The states consider both point and nonpoint sources of the constituent in their calculations.

The TMDL is then allocated to all of the point source dischargers, reserving some for the nonpoint sources. Some is also reserved for future discharges. Each of the point source discharges must stay within these limits. The TMDL is typically expressed as a load—for example, pounds of a constituent per day or kilograms of a constituent per day. This is calculated by multiplying the discharge flow rate times the constituent concentration. If the discharger increases the discharge flow rate at some time in the future, then the concentration of the constituent must be reduced to avoid exceeding the load. This frequently means additional treatment processes to improve removal.

Of the constituents in table 6.1, BOD is representative of the biodegradable organics that are discharged. Naturally occurring bacteria and other microorganisms in the river and stream use organics as a food source and consume oxygen in the process. This results in a reduction in the dissolved oxygen concentration in the water body and can cause the fish to suffocate (fish kills). Excessive TSS in the effluent causes cloudiness in the water and interferes with disinfection. The pH is limited to ensure the water remains in a near-neutral condition, avoiding excessively acidic or caustic conditions. Under all conditions, a minimum of 85 percent removal of BOD and TSS is required.

In addition to the constituents in table 6.1, limits are frequently placed on other pollutants, depending on the beneficial uses of the receiving water body. Those that serve as municipal drinking-water supplies have the most stringent requirements. Nutrients, such as nitrogen and phosphorus, are often limited, since they promote the growth of algae and plants in rivers and lakes. Ammonia, common in many wastewater effluents, is very toxic to fish and aquatic life. Like organic matter, it is also oxidized to nitrate by naturally occurring bacteria in the water consuming large amounts of oxygen. This reduces the amount of dissolved oxygen in the water and can cause fish to suffocate.

Phosphorus is present in all municipal wastewaters to some degree; but large amounts of phosphorus are associated with sediments washed off heavily fertilized fields. In the past, wastewater contained substantial amounts of phosphorus principally from detergents, but high-phosphorus detergents have been banned in many states and localities. As a result, the phosphorus concentration in wastewater has been reduced. Phosphorus is an essential nutrient for algae and plant growth. As a result, in many water bodies, algae must be carefully controlled by limiting the amount of nutrients discharged.

Groundwater

Maintaining the quality of groundwater is left up to the states' water pollution control agencies. Groundwater is an important source of water all over the world. In 2005, 23 percent of fresh water used in the United States was from groundwater; about one-fifth of that is for drinking and public water supply; the bulk of the remainder is for agricultural irrigation (Kenny 2009). For communities and industries located away from a water body, the percolation and evaporation of treated wastewater effluent are the only alternatives for disposal. If the treated effluent is percolated, it will eventually reach the groundwater. The constituents frequently regulated in groundwater discharges are salinity (total dissolved solids or mineral content), chlorides, nitrogen, and toxics. Nitrogen is particularly problematic because it is costly to remove and is regulated at a maximum of 45 mg/L as nitrate in drinking water. Excessive nitrate in drinking water causes methemoglobinemia (blue-baby syndrome) in infants. This is caused by decreased oxygen carrying capacity of blood in babies resulting from the use of nitrate carrying water for reconstituting the baby formula.

In areas where groundwater is used as a water supply, care must be taken to ensure the quality is not degraded. In some cases, only the evaporation of the wastewater effluent is allowed (no percolation). In these cases, any open evaporation ponds must be lined with a rubber/plastic membrane sealed to prevent leakage. In critical locations, multiple liners are used.

Disinfection and Control of Waterborne Diseases

Early history showed a link between inadequate sanitation and waste disposal and epidemics of waterborne diseases, such as cholera, dysentery, and typhoid. Municipal wastewater contains many of these infective agents, or pathogens. Pathogenic microorganisms come in several different forms, including bacteria, protozoans, helminths (worms), and viruses. These pathogens originate from the intestinal tracts of humans and animals and enter the human body through oral ingestion (Metcalf and Eddy 2003). Though they are not present in large numbers in wastewater, they must be killed off or otherwise inactivated so they are not able to cause illness. The term *disinfection* is used to describe this process of killing or inactivating pathogens.

Because the numbers of pathogens in wastewater are small and the varieties of possible pathogens are numerous, it is impractical to test selectively for all of these pathogens individually in treated wastewater effluent. So, in place of testing for all of them individually, an indicator organism is used. The indicator organism should be present when pathogens are present and absent when pathogens are absent. The indicator organism should come from the same source as the pathogen—that is, the enteric tract of humans and animals—and should respond in the environment and to disinfecting agents similar to the pathogens.

The indicator organism should be easy to identify. Ideally it should be present in very large numbers compared to the pathogens. (Why? Because if the number of indicator organisms is small in the treated effluent, and they outnumber the pathogens by hundreds of thousands in the untreated sewage, it can be assured the number of pathogens in the treated effluent must be very, very small.) Although no organism fits all of the indicator organism characteristics perfectly, the coliform organism matches most of them and has been the standard for measuring disinfection.

There are many different species of coliforms, a number of which are not necessarily associated with the intestinal tract of humans and animals. One subspecies of coliforms are fecal coliforms, which are more likely to inhabit the intestinal tract. One such subspecies is *E. coli*, which is also used as an indicator organism for that reason.

Regulatory agencies set limits on the maximum concentration of total coliform, fecal coliform, *E. coli* and other organisms depending on the uses of the river or lake. The limits vary from location to location.

Toxicity Testing

Discharges from wastewater treatment plants must not be toxic to fish and other aquatic species. The EPA has developed toxicity testing protocols to determine if wastewater effluents are potentially toxic to aquatic life (USEPA, Whole Effluent Toxicity, 2013). Often a single contaminant of a given concentration may not be toxic; but in the presence of other compounds or pollutants, even in small concentrations, it may be toxic. In other words, there could be a synergistic effect. With the myriad of chemicals already manufactured and the number increasing almost daily, it is not practical or even cost-effective to test for these compounds individually. And then it would still not be possible to find out if there was a synergistic effect.

To determine if the effluent from a wastewater treatment plant is toxic, toxicity testing of the effluent is typically required for discharges to fresh and ocean waters. These tests measure the acute toxicity (response, typically death, within a short period of exposure) or chronic toxicity (more long-term changes in the species) to selected species, such as minnows, algae, and kelp.

Recycled Water

The quality of treated effluent to be recycled is regulated by the states. The quality depends on the recycled water use. (This is discussed in chapter 9.)

Constituents of Emerging Concern

Many of the chemicals, personal care products, and pharmaceuticals that we use every day wind up in the wastewater we generate. Many of these chemicals are endocrine disruptors, which may interfere with the body's endocrine system and produce adverse developmental, reproductive, neurological, and immune effects in both humans and wildlife. In addition to pharmaceuticals and personal care products, dioxin and dioxin-like compounds, polychlorinated biphenyls, DDT and other pesticides, and plasticizers act as endocrine disruptors (NIEHS n.d.). Treatment plants do not remove these compounds entirely; most are only partially removed. As a result, these are discharged and eventually reach surface and groundwater. Although we should be concerned, there is no reason to be alarmed. The concentrations are in the part-per-trillion range. To put this in perspective on the basis of 1 mL = 20 drops or 0.05 grams per drop:

- 1 part per million ≈ 1 drop of water in 13 gallons of water
- 1 part per billion ≈ 1 drop of water in 13,000 gallons (one residential swimming pool)
- 1 part per trillion ≈ 1 drop of water in 13 million gallons (1,000 residential swimming pools)

Advanced treatment processes are used to remove these constituents in critical applications. These include injecting treated wastewater into a ground water aquifer which may subsequently be tapped as a drinking water supply source or discharging into a water reservoir used as a source of drinking water with or without any additional treatment (known as indirect potable water reuse).

Implementation and Enforcement

Each point-source discharger must have a permit to discharge to either surface water or groundwater. The permits specify the effluent limits and a monitoring and reporting program.

The dischargers provide self-monitoring under penalty of perjury. They are subject to significant fines and penalties for permit or reporting violations. Fabricated or falsified data reporting can lead to criminal prosecution, severe fines, and possibly imprisonment. The permits allow the regulatory agencies access to the facilities for sampling and inspection, unannounced, at any time.

References

33 U.S.C., 1251 et. seq. 2002. "Federal Water Pollution Control Ac t[As Amended Through P.L. 107–303, November 27, 2002]." November 27.

ASCE. 2013. *2013 Report Card for America's Infrastructure* . Accessed November 2, 2013. http://www.infrastructurereportcard.org/a/#p/wastewater/conditions-and-capacity.

Burian, Steven J., Stephan J. Nix, Robert E. Pitt, and S. Rocky Durrans. 2000. "Urban Wastewater Management in the United States: Past Present and Future." *Journal of Urban Technology* 7 (3): 33-62.

Foil, J. L., J. A. Cerwick, and J. E. White. 1993. "Collection Systems Past and Present, A Historical Perspective of Design, Operation and Maintenance." *Operations Forum*, December.

Griswold, Eliza. 2012. "How 'Silent Spring' Ignited the Environmental Movement." *New York Times Magazine.* Edited by Sheila Glaser. New York, September 21.

Kenny, J.F., Barber, N. L., Hutson, S.S, Linsey K. S., Lovelace, J. K., Maupin, M.A. 2009. "Estiamted Water Use in the United States in 2005." USGS Circular: 1344.

Metcalf, and Eddy. 2003. *Wastewater Engineering Treatment and Reuse.* 4th . Edited by G., Burton, F. L., and Stensel, H. D. Tchoganoglous. Boston, MA: McGraw Hill.

NIEHS. n.d. National Institute for Environmental Health Sciences. Accessed March 2, 2014. http://www.niehs.nih.gov/health/topics/agents/endocrine/.

NOAA. 2008. *Non-Point Source Pollution, Categories of Pollution: Point Sources.* March 25. Accessed September 20, 2014. http://oceanservice.noaa.gov/education/kits/pollution/03pointsource.html.

—. 2008. *Welcome to Non-Point Source Pollution, Non-point Source Pollution Agricultural Operations.* March 25. Accessed September 20, 2014. http://oceanservice.noaa.gov/education/kits/pollution/06operations.html.

Scott, Michael. 2009. "Cuyahoga River fire 40 years ago ignited an ongoing cleanup campaign." *The Plain Dealer.* Cleveland, June 22.

Service, Action Septic Tank. n.d. Accessed November 3, 2013. http://www.actionseptictankservice.com/.

USEPA. 2010. *NPDES Permit Writer's Manual.* USEPA .

—. 2013. *Whole Effluent Toxicity.* September 12. Accessed May 23, 2015. http://water.epa.gov/scitech/methods/cwa/wet/.

Chapter 7

Wastewater Treatment Stages

Introduction

Wastewater, as pointed out in chapter 4, is essentially 99.95 percent water by weight. The remaining 0.05 percent is material dissolved or suspended in water, which makes it wastewater. The pollutants that wastewater treatment plants are designed to remove are in this small 0.05 percent fraction of wastewater. These pollutants are made up of both organic and inorganic materials with their origin largely in the food that we consume and the materials we "dump" down our sewers, including animals and their waste products, street washings, rainwater runoff, and other activities. In separate sanitary sewer systems, animal droppings, street washings, and rainfall runoff would generally not be present.

Organic compounds are largely composed of a combination of carbon, hydrogen, oxygen, and nitrogen. Other important elements, such as sulfur, phosphorus, and iron, may also be present. All these elements are contributed by the food we consume via proteins, carbohydrates, fats, and oils. (This was discussed in chapter 4.)

To remove these and other pollutants, wastewater treatment plants use several stages of treatment. These are generally classified into the following major categories:

1. preliminary treatment
2. primary treatment
3. secondary treatment
4. tertiary treatment
5. advanced treatment

Each stage may include one or more processes or operations. In addition, each step or category produces a side stream (waste stream) that contains the materials that were removed in the process. This waste stream is called reject water, recycled flow, biosolids, or sludge, depending on its source and composition. Side streams are handled by another set of processes, which are presented in chapter 8. This chapter discusses only processes for handling/treating liquid portion of wastewater. (See figure 7.1, a diagram showing a typical liquid processing system.)

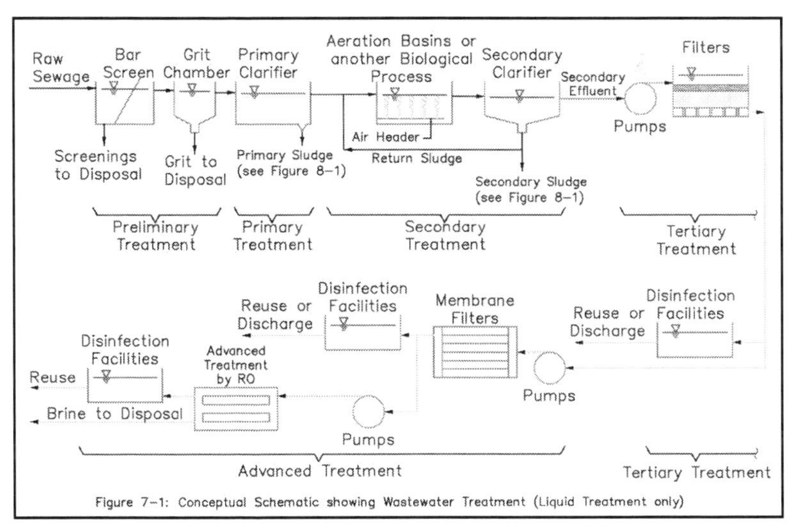

Figure 7-1: Conceptual Schematic showing Wastewater Treatment (Liquid Treatment only)

Figure 7.1 shows the five stages of treatment listed above.

Preliminary Treatment

Preliminary treatment usually consists of screening and grit removal. This usually takes place in a part of the treatment plant called the headworks. This portion of the plant is the most malodorous and, as such, is usually totally enclosed. Malodorous air is collected and treated using special processes to remove the odors before releasing the air into the atmosphere.

As a general rule, preliminary treatment does not significantly reduce any of the important quality parameters, such as BOD, COD, or TSS. Rather, it protects equipment in follow-on processes from this abrasive and clogging material. Despite having a minimal impact on BOD, TSS, etc., preliminary treatment is a very important step in any treatment plant.

Typical Wastewater Screens

Screening. The removal of large solids that could damage downstream process elements is usually achieved by passing the waste through a channel containing a screen. Solids (rags, "flushable" wipes, wood, toys, stones, large objects, etc.) that collect on the screen can be removed by automatic raking mechanisms or manual raking for separate disposal, or they can be ground up and removed or returned to the waste stream.

The trend in waste treatment today is to install mechanically raked screens with few moving submerged parts. This removes the debris from the plant and reduces cleaning chores in downstream units. The collected screenings are commonly washed to minimize odors and compacted for disposal in a landfill.

Grit removal. The removal of grit after screening is required because grit is very abrasive and can increase wear on downstream piping, pumping, and treatment units. Grit consists of relatively inert and relatively nonbiodegradable materials, such as sand, gravel, cinder, egg shells, coffee grounds, and other heavy materials that settle readily by gravity. If not removed in the preliminary treatment, these materials will settle in subsequent processes and become part of the sludge, ultimately taking up valuable space in the solids processing system (described in chapter 8).

The removal of grit is accomplished in grit chambers, which typically are small tanks (with a detention time of two to five minutes) where grit settles by gravity and is removed by special grit pumps to a grit washer system designed to remove the organics, dewater the grit, and render the grit less odorous. There are many different types of grit removal systems, including long, narrow channels, proprietary vortex-type systems, and aerated grit chambers. Channel and vortex systems rely on velocity to keep organics out of the grit. In an aerated grit chamber, compressed air is blown into the bottom of the tank, where it keeps lightweight organic materials in suspension while allowing the grit to settle. The grit is ultimately hauled away to a landfill.

Raw wastewater pumping. The sewers coming into a treatment plant are often deep in the ground. The wastewater must be pumped up to ground level or slightly above to allow the water to flow through the treatment processes as much as possible by gravity. Pumping facilities can be located before or after the screening and grit removal; most of the time the pumping system is before screening and grit removal. Special clog-resistant pumps are used, since they must pump whatever is "dumped" into the sewer.

Modern Wastewater Pumping Station

Primary Treatment

Empty Primary Clarifier

Primary treatment consists of passing the wastewater through large settling basins, or "clarifiers" (detention time, two to three hours), allowing those solids that will settle under the influence of gravity to separate from the bulk of the wastewater. In addition, oil and grease are removed in the primary settling process. The quiescent water permits the oil and grease to float to the top, where it can be skimmed off with mechanical skimmers. That material is called scum and is pumped to a sludge treatment system.

The solids, which settle to the bottom of the primary clarifier, are removed by mechanical scrapers/collectors and conveyed to a solids treatment system as primary sludge. The primary treatment process removes approximately 30 to 35 percent of the raw wastewater BOD and approximately 60 percent of the suspended solids without a significant consumption of energy. The removal percentage of BOD and TSS can be improved through the addition of chemicals; this increases the operating cost, but can be a quick stop-gap measure to improve performance and reduce the organic and solids load to the secondary treatment process.

Primary clarifiers are often covered to control odors.

Primary clarifiers or primary sedimentation tanks can be circular, square, or rectangular. As a general rule, rectangular or square designs require less space, whereas circular designs are considered to require less maintenance but take up more space. The circular units use scraper mechanisms that are more reliable than the plastic chains used in rectangular clarifiers, which are subject to breaking.

Primary clarifiers can be a source of odors and hence are often covered completely. The tanks are ventilated through odor scrubbers. Primary clarifiers are rarely used in smaller wastewater treatment plants in order to reduce costs and simplify operations; the screened and degritted wastewater flows directly to the secondary treatment process.

Secondary Treatment

Many processes have been developed over the years, any one of which by itself or in combination with others can provide secondary treatment. Secondary treatment processes are normally biological processes that culture and concentrate microorganisms to use the organic matter remaining in the wastewater as a food source.

Broadly, these processes (1) employ a fixed media—such as rocks, PVC bundles, wood, or gravel with large surface area for microorganisms to grow on—installed in a structure; (2) tanks where the microorganisms are kept suspended in wastewater mechanically or with air, which provides agitation as well as oxygen; or (3) some combination of the two. These are generically classified as fixed film systems, suspended growth systems, or hybrid/integrated systems that harbor the microorganisms.

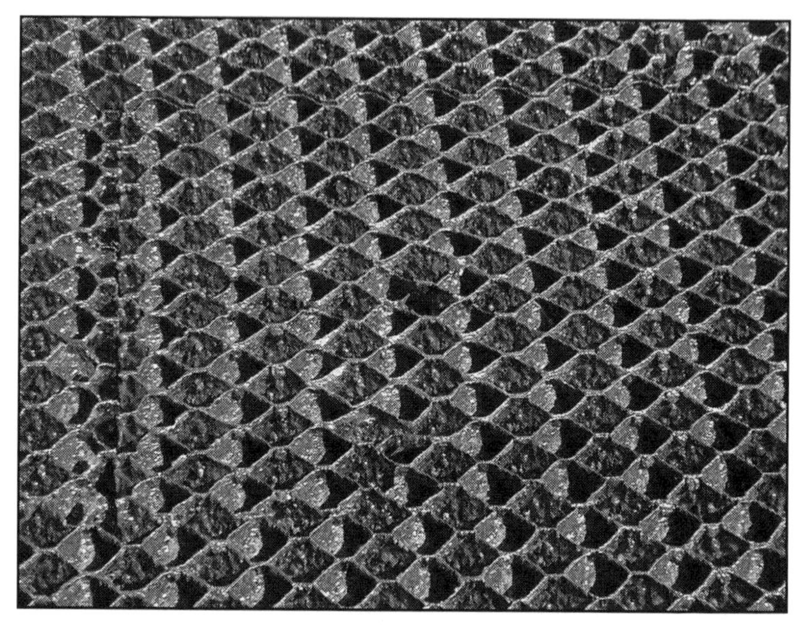

Biofilm on Trickling Filter Media

The major purpose of secondary treatment is to remove the soluble BOD that escapes primary treatment and to provide further removal of the very small, suspended solids that are not removed in the primary treatment process. As mentioned previously, at a minimum, secondary treatment is now required for all the plants in the United States unless a special waiver is approved by the US Environmental Protection Agency. Several enhancements and modifications of these processes have been developed to achieve project-specific goals, such as the removal of nutrients like nitrogen and phosphorous, where the plant effluent will be discharged into lakes and reservoirs where algal blooms by nutrient inputs are major concerns.

Secondary treatment processes are designed to provide the proper environment for the biological breakdown of soluble organic materials in wastewater. A great variety of biological microorganisms come into play and interact: bacteria, protozoa, rotifers, fungi, algae, and others. Under the proper conditions of reaction time, oxygen, temperature, pH, suitable nutrients, and an environment free of toxins, the microorganisms biodegrade organics into carbon dioxide, water, and other minor end products and produce more cell material (more microorganisms). Generally, the larger the number of microorganisms that can be grown in a tank, the greater its treatment capacity. The microorganisms are then settled in secondary clarifiers, which separate the microorganisms, leaving a relatively clear liquid called secondary effluent.

Trickling Filter

There are two common secondary processes: activated sludge and trickling filters (biotowers). In the activated sludge process, the microorganisms are suspended in the wastewater in large tanks. Compressed air is bubbled in the bottom of the tanks or added with "splashers" at the surface. This is necessary to provide oxygen to the hungry microorganisms. The liquid in the tanks is a brownish color with an earthy smell. If the color is black or there are other odors, there is something wrong.

A trickling filter (biotower) uses rocks or plastic media in a reinforced concrete structure as a place for microorganisms to grow. Trickling filters, considered older technology, have a rock media depth of about six feet; biotowers have plastic media twenty to thirty feet deep. A liquid distributor rotates above the media and uniformly distributes the primary effluent on the rock or plastic media, and the microorganisms extract the organics from the trickling liquid. Many of these systems include a method to recirculate some of the effluent back through the process to improve operation.

Aeration Tank
out of Service

Activated Sludge Aeration
Tank in Operation

Integrated or hybrid systems are coming into use. These systems have plastic elements suspended in the aeration tanks of the conventional activated sludge process. The plastic elements provide a growth area for additional microorganisms and as such can provide more microorganisms per tank than otherwise possible. This permits a greater treatment capacity with less tank volume.

Typical Activated Sludge

Up until now, the discussion has been limited to the removal of BOD. Many plants that discharge to surface waters must remove ammonia, which is toxic to aquatic life (see chapter 4). Ammonia can be biochemically oxidized to the nontoxic nitrate form (called nitrification) by a certain species of bacteria (nitrifying bacteria). These bacteria live naturally in soil and water, and so they can be grown in the secondary treatment system.

They are, however, slow growing and need time to multiply and accomplish nitrification. In the activated sludge system, this is accomplished by increasing the time the microorganisms are kept alive in the system. To accomplish BOD removal, the solids retention time (residence time in the process or SRT) can be as short as a day or two; but is generally kept closer to four to five days or so to create a more stable operating system. To be able to grow the nitrifying bacteria, six to ten days or more of solids residence time (SRT) are needed, depending on the temperature of the wastewater liquid. Providing this additional time in the system requires larger tanks and consequently more cost. In addition, the nitrification process requires much more oxygen than BOD removal.

Secondary clarification. Secondary clarifiers, like primary clarifiers, can be rectangular or circular (the two most common) or square, and they follow the aeration tanks, trickling filters (biotowers), or other biological process. A retention time of about three to four hours is provided for the sludge or biomass to settle. In the activated sludge process, the biomass is separated from the liquid and concentrates at the bottom of the secondary clarifiers. The concentrated biomass, which has been starved of food for a few hours, is pumped back to the aeration tank to consume more incoming wastewater organics. The return solids are called return activated sludge. It is important to note that in the process of consuming organics and under the proper environment, the microorganisms reproduce. And as long as there is a food source (wastewater organics) and the proper environment, they will continue to grow. If nothing is done, they will grow to the point where the secondary clarifier will not be able to separate them from the liquid any more. If no separation occurs, the excess microorganisms will be discharged with the effluent—a most undesirable situation! So to prevent this, a certain amount of solids is removed from the process each day. This amount can be calculated based on the BOD coming into the plant and the amount of growth anticipated. The solids that are wasted, called waste activated sludge, are sent to the sludge-handling system of the treatment plant. In a trickling filter (biotower) treatment process, the settled solids are generally not recycled back to the filter.

Typical Secondary Clarifier

The liquid portion from the secondary clarifiers, known as secondary effluent, can be disinfected to kill or inactivate a major portion of any remaining pathogens, microorganisms, and protozoa if no further treatment is required for discharge or reuse. However, some reuse and/or discharge situations require further treatment, as discussed in chapter 9. In such cases, additional treatment is required.

The secondary treatment process can be configured to provide nitrogen reduction through a process called biological denitrification or to achieve biological phosphorus reduction. To implement these systems, the tank design and the environment in the tank are modified. Biological denitrification can reduce the total nitrogen to about 4 to 5 mg/L or even less; enhanced biological phosphorus removal can reduce total phosphorus to about 2 mg/L (USEPA, 1987) (USEPA, Nutrient Control Design Manual, 2010). Reducing phosphorus below 2 mg/L requires chemical precipitation as part of the primary treatment or tertiary treatment.

Tertiary Treatment

Although secondary treatment processes, when coupled with disinfection, can remove more than 85 percent of the BOD and suspended solids and nearly all pathogens, there is still a small amount of suspended matter, suspended organics, phosphorus, nitrogen, and dissolved, nonbiodegradable organics remaining. Where these pollutants are a major concern, which depends on the type of reuse intended, there are processes capable of removing them. They come in the category of tertiary or advanced treatment.

Treated Wastewater Ready for Reuse

Tertiary treatment, in most cases, is comprised of filtration through a granular media (typically sand and/or anthracite) with a bed thickness of about three to five feet; disk filters made of a synthetic material such as polyester or nylon fiber; or microfiltration using polypropylene membranes or some other configuration. These filtration systems require periodic cleaning (backwashing), which is a "waste stream" containing the solids that were removed and that require further treatment. Typically this waste stream is returned upstream to the preliminary treatment process.

To improve the efficiency of removal for filters, chemicals such as aluminum sulfate and polymers are often added in small quantities.

These processes improve effluent quality to the point that it is adequate for many reuse purposes, converting a wastewater into a valuable resource.

Advanced Treatment

Even after tertiary treatment, the soluble organic materials and chemicals of emerging concern that are resistant to biological breakdown persist in the effluent. The persistent materials are commonly referred to as refractory organics. Secondary effluent COD values are often 30 to 60 mg/L with a BOD of about 20 to 30 mg/L.

The removal of these recalcitrant and refractory (treatment-resisting) organics that remain after tertiary treatment is becoming increasingly important in many reuse applications intending to supplement, indirectly, the potable water supply through groundwater percolation or injection. This is generally accomplished by another process involving membranes with very small pores, known as reverse osmosis. These membranes selectively remove desired constituents, including both the refractory organics as well as the dissolved minerals (sodium, potassium, magnesium, silica, etc.) naturally present in water and therefore the wastewater. After this treatment, the water is as good as, and perhaps better than, drinking water in most respects.

In many cases, high dosage of ultraviolet irradiation, in conjunction with an oxidant, like hydrogen peroxide, is used following reverse osmosis to further remove organics and achieve or enhance disinfection, i.e., pathogen kill.

Although, after this treatment, the water is essentially free of pathogens, viruses, protozoa, and other disease-causing organisms, it receives another level of treatment known as disinfection to kill any remaining disease-causing organisms to be absolutely certain that it is safe for all uses, including indirect or direct potable reuse via ground water injection, discharge into drinking water reservoirs and other uses where there may be lingering health concerns about its use. Disinfection is accomplished by chlorination, ozonation, ultraviolet radiation (UV), or a combination of UV and hydrogen peroxide.

Advanced treatment is expensive. It has a high capital cost and a high operational cost resulting from high usages of chemicals and power. In addition, the processes or process combinations have a relatively short history of successful use. Generally, pilot studies are run for a year or so to optimize their design and develop design criteria aimed at achieving reliability, cost savings, and ease of operation.

References

USEPA. (1987). *Design Manual Phosphorus Removal.* Cincinnatti, OH, USA: Water Engineering Research Laboratory.

USEPA. (2010). *Nutrient Control Design Manual.* Cincinnati, OH, USA: Office of Research and Development/National Risk Management Research Laboratory.

Chapter 8

Biosolids Treatment and Handling

Introduction

As discussed in chapter 7, biosolids, sometimes called sludge, are the byproducts of wastewater treatment. Technically speaking, biosolids are wastewater sludges that have undergone a stabilization process to permit the solids to be beneficially reused. Sludges are principally produced in the primary and secondary treatment processes and consist of suspended matter in sewage (called primary sludge in primary treatment) and excess microorganisms/biomass combined with remaining suspended matter (called secondary sludge in secondary treatment). Together, about 1,500 to 2,000 pounds of biosolids, by dry weight, are produced from the treatment of 1 million gallons of domestic wastewater, of which about 60 percent is primary sludge and the remaining 40 percent is secondary sludge. These percentages and quantities are approximate and depend largely on wastewater characteristics and the type of treatment provided at the plant. Primary sludge is usually thicker than secondary sludge—typically 3 percent to 6 percent dry solids or more. Secondary sludge from a trickling filter (biotower) is 1.5 percent to 3 percent dry solids; and the waste solids from an activated sludge clarifier is usually less than 1 percent.

Biosolids Drying Beds

Since the sludge is not dry (that is, 100 percent solids and no water) but is 95 to 99 percent water, with the remaining 1 to 5 percent solids, the quantities of sludge are many times larger. For example, if the combined sludge (primary plus secondary) totals 2,000 pounds on a dry-weight basis and has 2 percent solids in it, the total quantity of sludge will be (100 divided by 2) times 2,000 pounds—that is, 50 times 2,000 pounds or 100,000 pounds of "wet sludge," or a volume of approximately 12,000 gallons. This is for treating only 1 million gallons of wastewater.

If the sludge is only 1 percent solids, the quantity of wet sludge will be twice as much (24,000 gallons). Just as a reference, 1 million gallons a day of wastewater (1 mgd) is generated by about 10,000 to 12,000 people. So for a community of 100,000 people, the total wet sludge generated as a result of primary plus secondary treatment could be as much as 100,000 to 120,000 gallons per day, assuming it contains 2 percent solids. Handling such a large quantity of sludge, which is a byproduct of wastewater treatment, poses an equally difficult challenge in the treatment of wastewater.

Before the Clean Water Act in 1972, when there were no strict regulations for the disposal of sludge, communities would discharge it into the ocean, dump it in landfills, surface apply it on land, and use it for agriculture, silviculture (forests and tree farms), and other similar unregulated disposal sites without any treatment. As a result, it would create unsightly and unhealthy conditions, attracting rodents, mosquitoes, and birds spreading a host of pathogens. For these reasons, thankfully, this practice is no longer allowed under new laws and regulations that have been passed by Congress, state health authorities, and local jurisdictions. Sometimes local laws are more stringent than federal and state laws due to pressure from local citizenry and can prevail unless successfully challenged in court.

Objectives of Sludge Handling

Figure 8.1 shows the typical steps in processing sludge from treating wastewater and the subsequent energy recovery that is possible.

Today all the processes for sludge treatment at wastewater treatment plants have several objectives in most cases: (1) stabilize the sludge to reduce pathogens and vector attraction (birds, rodents, etc.) and reduce the quantity of dry solids to be disposed; (2) thicken and dewater the sludge so that it gets concentrated to, say, 20 to 25 percent solids, thus reducing the total quantity of sludge by about 90 percent (to about 1,200 or fewer gallons per million gallons of wastewater); and (3) if possible, produce digester gas and power by anaerobic digestion so that the treatment plant's operation is sustainable with a reduced carbon footprint. (Note that in small plants of less than 5 million gallons per day, anaerobic digestion may not be cost-effective.) When the sludge is concentrated to 25 percent solids or so, it can be transported easily in trucks.

The thickening of primary sludge is usually not required, as there is not much benefit if the concentration is already close to 5 or 6 percent; but the thickening of secondary sludges, particularly waste activated sludge, is essential to minimizing the size and cost of sludge handling systems.

Sludge thickening is done using different technologies (gravity thickeners, gravity belt thickeners, flotation thickeners, centrifuges, etc.) that thicken it from its initial concentration of 2 percent or less to 4 to 7 percent. At this concentration level, the thickened sludge is fed to anaerobic digesters for digestion and gas production. The digesters are egg-shaped or cylindrical structures—heated and mixed—with a detention time of fifteen to twenty days or more. During digestion, a significant portion of volatile solids is destroyed (from 35 to 50 percent), producing digester gas.

As a result of digestion, the solids content of the sludge is reduced (but not the liquid content), thus reducing its concentration accordingly. The digested sludge is then dewatered using another technology (centrifuges, screw presses, belt filter presses, etc.), which further concentrates the sludge to 20 to 25 percent solids or more, thereby significantly reducing the quantity of sludge by another 80 to 85 percent, which subsequent processes must handle.

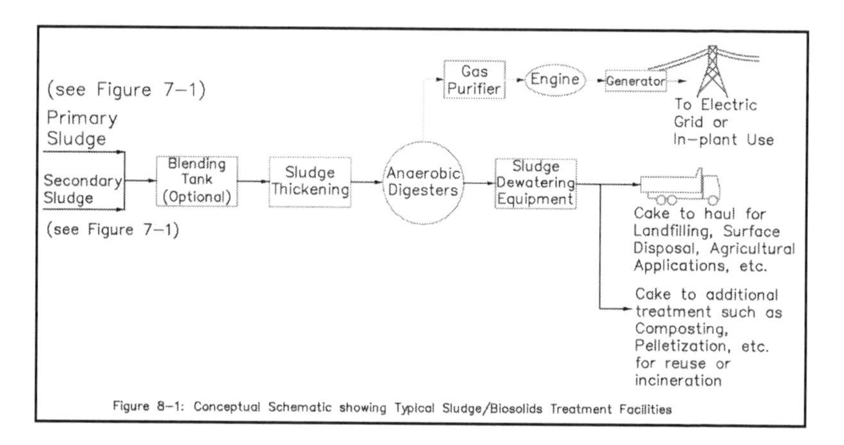

Figure 8–1: Conceptual Schematic showing Typical Sludge/Biosolids Treatment Facilities

With all processes combined, the quantity of sludge has been reduced from 12,000 gallons per 1 million gallons of wastewater at 2 percent concentration in the beginning to about 100 cubic feet, or about 800 gallons. It is now much easier to handle than the initial quantity of 12,000 gallons. At this concentration, the sludge has a consistency of toothpaste and is called sludge cake.

Once the sludge has gone through the digestion or stabilization process, it can be called biosolids and put to beneficial use. The permitted uses depend on the degree of stabilization and the processes used.

Egg-Shaped Anaerobic Digesters

Conventional Cylindrical Anaerobic Digester

Depending on the temperature maintained in the digesters, biosolids (and biosolid cake) are classified as Class A or Class B. A temperature of about 130 degrees F (thermophilic digestion) produces Class A biosolids, while a temperature of about 95 degrees F (mesophilic digestion) produces Class B sludge. The higher temperature in thermophilic digestion reduces the pathogen content of the sludge beyond what the lower temperature in mesophilic digestion does. In addition, the higher temperature further reduces the volatile content of the sludge, thus minimizing rodent and vector nuisance. Each type of biosolid has associated requirements and restrictions for its reuse specified by the US Environmental Protection Agency (EPA) (40 CFR 503 Sludge Regulations), with more restrictions for Class B sludge as compared to Class A sludge.

Sludge Regulations

The Clean Water Act, as enacted in 1972, addressed sewage sludge use and disposal in only one limited circumstance: when the use or disposal posed a threat to navigable waters.

Thus, Section 405(a) of the act prohibited the disposal of sludge if it would result in any pollutant from the sludge entering navigable waters unless it complied with a permit issued by the EPA. In 1977, Congress amended Section 405 to add Section 405(d), which required the EPA to develop regulations containing guidelines for the use and disposal of sludge. These guidelines were to (1) identify uses for sludge, including disposal; (2) specify factors to be taken into account in determining the methods and practices applicable to much of these identified uses; and (3) identify concentrations of pollutants that would interfere with each use.

In 1987, Congress amended Section 405 and for the first time set forth a comprehensive program for reducing the potential environmental risks and maximizing the beneficial use of sludge. The amended Section 405(d) established a timetable for the development of the sewage sludge use and disposal guidelines. As the basis of the program, Congress mandated the development of technical standards for sewage sludge use and disposal to protect public health and the environment, and the implementation of these standards, in part, through a permit program.

Under Section 405(d), the EPA was mandated to first identify, based on available information, toxic pollutants that may be present in sewage sludge in concentrations that may affect public health and the environment. Next, for each identified use or disposal method, the EPA was required to promulgate regulations that specified acceptable management practices and numerical limitations for sludge that contains these pollutants. These regulations were to be "adequate to protect human health and the environment from any reasonably anticipated adverse effect of each pollutant." The statute required the EPA to promulgate sewage sludge regulations in phases and periodically to review these regulations to identify additional toxic pollutants for regulation.

To develop these standards, the agency conducted the National Sewage Sludge Survey (NSSS), which gathered, among other things, additional information on the pollutants in sewage sludge, how sludge is used and disposed, and information on the management of sludge. In cooperation with other agency offices and outside expert reviewers, the EPA also gathered data on the movement of certain pollutants into and through the environment; refined and expanded its modeling capability for specific pollutants or disposal practices (for example, surface disposal sites); supplemented its information on other disposal practices (sewage sludge incinerators); and further examined the characteristics of partially treated sewage or sludge in septic tanks, also known as septage.

All these efforts culminated in the development and signing of these standards (503 sludge regulations) on November 25, 1992, by the EPA. These regulations became a law after they were published in the Federal Register, which occurred in 1993. The regulations are continually updated as additional data becomes available, both on the pollutants present in wastewater sludges as well as on their effects on human health and the environment.

Summary of the Final Rule

The 503 rule established standards for the final use or disposal of sewage sludge when the sewage sludge is applied to agricultural and nonagricultural land (including sewage sludge and sewage sludge products sold or given away—described in the proposed rule as distributed and marketed sludge), placed in or on surface disposal sites, or incinerated. The rule does not apply to the processing of sewage sludge before its ultimate use or disposal. In addition, in this rule, the EPA did not specify process operating methods or requirements for sludge entering or leaving a particular treatment process. In other words, the end is specified, not the means to that end.

The regulations include specific numerical limits for ten pollutants when sewage sludge is used or disposed of by the above methods:

1. arsenic
2. cadmium
3. chromium
4. copper
5. lead
6. mercury
7. molybdenum
8. nickel
9. selenium
10. zinc

The reason for limiting these heavy-metal concentrations was fear that if these were applied to the land in excessive quantities, plants would uptake these metals and then enter the food chain, either directly in fruits and vegetables or in the fodder and grass that cattle and other farm animals would eat. The maximum concentration of these metals in terms of kilograms of metal per kilogram of dry sludge solids as well as the annual and total application rates per unit area (per acre or hectare) are specified in the final rule.

Supplementing the numerical pollutant limits are management practices and general requirements to protect human health and to prevent the gross abuse of the environment. In the case of the small quantity of sludge that is sold or given away in a bag or other container, the rule requires the treatment works (or person, if different from the treatment works) to label the product. The label is to provide instructions on properly using the product. This type of sludge must be Class A.

The rule also includes monitoring, recordkeeping, and reporting requirements. The frequency with which sewage sludge is to be monitored depends on the quantity of sludge used or disposed of by a treatment plant. Similarly, the pollutants that treatment plants must monitor their sewage sludge for depend on the use or disposal method selected. The recordkeeping and reporting requirements are also specific for each method of use or disposal.

So, in essence, the 503 rule applies to only the following practices:

- land disposal of sludge, including application of sludge to agricultural and nonagricultural land (including sewage sludge or sewage sludge products sold or given away);

- surface disposal of sludge, including sludge-only landfills, often referred to as monofills, and disposal on dedicated sites with or without a vegetative cover;

- incineration of sludge (with no more than 30 percent by weight of municipal solid waste as an auxiliary fuel); and

- septage disposal.

The 503 rules do not apply to

- disposal of sludge in municipal landfills;

- disposal of industrial sludges;

- disposal of hazardous sludges;

- disposal of municipal wastewater treatment plant sludges by methods other than land disposal, surface disposal, and incineration as defined above.

Other regulations of the Clean Water Act, not 503 Regulations, apply to these "excluded" situations.

The good news is that National Sewage Sludge Survey (NSSS) determined that most municipalities are producing "clean sludge" (without excessive concentration of heavy metals) and are able to comply with the Rule 503 sludge use and disposal regulations without changing their management practices. Only the remaining 30 percent would be required to improve the quality of their sludges by industrial pretreatment or other means; use better sludge treatment methods to improve stabilization, pathogen kill, and vector control; and institute improved monitoring and recordkeeping.

Rule 503 aims at accomplishing sludge reuse; however, in doing so, it aims at protecting public health and safety as well as threatened and endangered species.

Bottom Line

The bottom line of these 503 Regulations is that the pre-1972 disposal of sludge on land, into the ocean, and indiscriminately in agriculture, horticulture, and silviculture is illegal. These regulations specify criteria for disposal of sludges from the treatment of domestic wastewater in terms of their pollutant contents, concentrations, application rates on land, surface disposal, management practices, recordkeeping, setbacks of disposal sites from property lines, and sampling frequencies. There are separate requirements for different disposal scenarios, the purpose of each being to accomplish sludge reuse—and, in doing so, to protect public health and safety as well as threatened and endangered species.

The 503 Rule is "self-implementing" in most cases. It is generally fully enforceable even in the absence of a permit, although the EPA has the legal mandate to include appropriate requirements of the rule in permits. The permits must include use and disposal requirements tailored to the characteristics of a particular project and site.

Digester Gas Utilization and Cogeneration

Digester gas contains about 60 to 65 percent methane (the gas we use in our homes for heating and cooking) and about 30 to 40 percent carbon dioxide. The heating value is about 600 BTU per cubic foot, or about two-thirds of that of natural gas. It has appreciable energy value.

The gas can be used to fuel a boiler to heat water, which can be used to heat the anaerobic digesters (remember, it needs to operate at 95 or 130 degrees F, depending on whether it is mesophilic or thermophilic digestion), or can be used to fuel an engine to drive a generator or a pump or compressor. A typical engine would turn only 30 to 40 percent of the energy value of the digester gas into real mechanical energy to drive either the pump or the generator. The other 60 to 70 percent of the gas energy would be wasted as heat, blown away with a radiator fan or out an exhaust stack.

To increase the efficiency from 30 to 40 percent to, say, 80 to 85 percent, heat generated in the process of power generation can be used to meet heating needs at the plant as well as to produce steam and more power using steam turbines, particularly in large plants. This system where both heat and power are produced with digester gas is called combined heat and power (CHP) or cogeneration. About 35 to 40 kilowatts of power can be produced by the sludge generated by 1 million gallons of domestic sewage without cogeneration and twice as much if excess heat is also used for this purpose.

Composting

Composting is a method of converting Class B or Unclassified biosolids to Class A biosolids. The composting process requires the sludge to be mixed with a bulking agent, such as wood chips or ground up "green waste" (yard and landscape waste). The mixture begins to heat up naturally due to the microbial action. Initially, the bacteria and microorganisms are mesophilic. The compost continues to heat up, and the microorganisms are replaced by the thermophilic type. After some time, there is little readily biodegradable material remaining. The compost goes through a cool-down period, and after a while a stable mature compost results. The compost can then be screened to remove the bulking agent, which can be reused.

A number of different technologies can be used, including (1) aerated static pile, (2) windrow, and (3) in-vessel. The aerated static pile is the simplest. Basically a pile of sludge and bulking agent, six to eight feet high, is created on the top of a series of perforated pipes. The pile is typically covered with screened and cured compost for insulation. A blower sucks air into and through the pile. The air is exhausted through a compost biofilter for odor control. Composting generally takes three to four weeks, followed by a thirty-day curing period.

In windrow composting, piles about six feet high and twelve to fifteen feet at the base are created. During the composting period, the piles are turned periodically with special motorized equipment. Aerobic conditions are not always maintainable, and sometimes when the piles are turned over, odors can result. The piles are typically not aerated, and require a turnover process. The compost period is three to four weeks (Metcalf & Eddy, 2003).

In-vessel composting takes place in a building or similar enclosed area. There are several different types of proprietary systems. Odors are easier to control with the in-vessel system.

Incineration

It is reported there are over two hundred municipal sludge incinerators operating in the United States (Earth Justice, 2014). The bulk of these are in the northeast and Great Lakes area. There are two types of incineration technology: multiple hearth and fluidized bed. Most are the multiple hearth type. Auxiliary fuel is typically used at least to get the furnace started. If the sludge is dry enough going in, it will combust without auxiliary fuel.

In a multiple hearth furnace, there are a series of hearths on a rotating center shaft. Air is blown in the bottom of the furnace, and the sludge is introduced at the top and drops onto the hearth below. As the hearth rotates, the sludge is pushed outward to the edge of the hearth, where it drops to the second hearth. As it rotates, sludge is pushed to the inside, toward the shaft, where it drops to the third hearth. This continues until the fully burned sludge ash drops out of the incinerator at the bottom.

In a fluidized bed furnace, a deep bed of sand is heated and kept fluidized by blowing air up through it. The sludge is introduced into the sand bed and combusted in the hot sand layer.

The ash that results from the combustion is blown out of the incinerator. This needs to be captured in air-pollution control equipment before being released into the atmosphere.

A sludge incinerator is a very costly investment and mechanically complex, requiring sophisticated operations. Generally incineration is practical only for large regional agencies. For estimating purposes, an incinerator should be considered only if the sludge generation rate is between 0.25 and 3 dry tons per hour (6 to 72 dry tons per day). Above 3 dry tons per hour might require multiple incinerators. Below 0.25 dry tons per hour, incineration may not be cost-effective. Permitting an incinerator in not an easy task. (NEIWPCC, 2001)

References

Earth Justice. (2014). *Cleaning Up Sewage Sludge Incinerators.* Retrieved May 17, 2014, from Earth Justice: http://earthjustice.org/our_work/cases/2013/cleaning-up-sewage-sludge-incinerators

Metcalf, & Eddy. (2003). *Wastewater Engineering Treatment and Reuse* (4th ed.). (G. B. Tchoganoglous, Ed.) Boston, MA, USA: McGraw Hill.

NEIWPCC. (2001, October). New England Interstate Water Pollution Control Commission Fact Sheet. *Sewage Sludge Incineration.* Lowell, MA, USA.

Chapter 9

Water Recycling and Reuse

Introduction

We pointed out in chapter 2 that water supplies are not unlimited, but the world population is increasing by the day. The quality of water in our raw water supplies, both underground and on the surface—such as in rivers, lakes and streams, and the oceans—is continually deteriorating. In addition, regulators are demanding better treatment of wastewaters as our ability to detect macro- and micropollutants is increasing with the development of new analytical tools. As communities grow and the available water supplies decrease, there is increasing emphasis on conservation. But conservation alone may not be enough to get through the next drought, so these communities look to water recycling as "drought insurance."

Next to conservation, and unless the community has an overabundance of fresh water, water recycling is typically the least costly source of water—generally much less expensive than desalination or importing water long distances and treating it for domestic use. As of 2012, thirty-two states have either guidelines or specific regulations relating to water recycling and reuse in place (USEPA, 2012). Although they differ somewhat from state to state, the purpose in all cases is the same: the protection of public health. Many states encourage water reuse and provide funding in the form of grants and low-interest loans for these types of projects. With increasing levels of treatment required for discharge, implementing water recycling should always be considered, because frequently little additional treatment is required to recycle the water.

Recycled Water Quality Requirements

Public Health

State guidelines or regulations for recycled water use relate to the level of treatment and disinfection for specific reuse applications. To ensure excellent disinfection where there is significant potential for public contact, particularly children, many states require tertiary filtration to remove as much of the suspended solids as possible so that pathogens are not able to shield themselves from the disinfecting agent—for example, chlorine or UV irradiation. Where access is restricted and public contact is minimal, disinfected secondary effluent can be used. Where recycled water is used for industrial cooling tower applications (evaporative cooling), drift eliminators must be in place, and biocides must be maintained in the circulating water to prevent legionella in the aerosols. Also, the air intakes for buildings should be carefully located to avoid drawing in aerosols from the towers.

Table 9.1 presents a summary of potential reuse applications and general water-quality requirements in terms of treatment and disinfection. The table is intended to be very general; each state may have different requirements.

Table 9.1
Reuse Applications and Suggested Treatment Requirements

Application	General Guideline for Recycled Water Quality (see notes below)
Landscape irrigation of public areas without restriction	TFD
Landscape irrigation of areas with restricted public access such as highway landscaping	DSE
Nurseries, Christmas tree farms, tree farms, and sod farms	DSE
Food crops	TFD
Food crops which undergo commercial, pathogen destroying processing before consumption	DSE
Pasture and fodder for animals, fiber (cotton) and seed crops not eaten by humans	DSE*
Orchards and vineyards not bearing food crops during irrigation	DSE
Supply for nonrestricted recreational impoundments	TFD
Supply for restricted recreational impoundments, fish hatcheries	TFD

Application	General Guideline for Recycled Water Quality (see notes below)
Toilet and urinal flushing in commercial buildings	TFD
Commercial and public laundries for clothes washing	TFD
Cooling tower makeup	TFD
Industrial process water (potential worker exposure)	TFD
Industrial process water (no worker exposure)	DSE
Boiler feed water	DSE
Firefighting	TFD
Firefighting by dumping from aircraft	DSE
Water jetting for backfill consolidation around potable water lines	TFD
Water jetting for backfill consolidation around other than potable water lines	DSE
Washing aggregate and making concrete	DSE
Dust control and fill moisture control	DSE
Decorative fountains	TFD
Snowmaking	TFD

TFD = Tertiary filtered and well disinfected effluent; total coliform < 2.2/100 mL
DSE = Disinfected secondary effluent; total coliform < 23/100 mL

* Disinfection may not be required depending on local regulation and practice.

In table 9.1, the method of applying water for irrigation—spray versus subsurface drip, for example—can make a difference in the level of treatment required. The public exposure to spray irrigation systems is much greater than to subsurface drip irrigation.

Even if the recycled water meets stringent public health requirements, there is always a concern for potential cross connections with public domestic water supplies. This is not much of a concern in public streets where the pipes are well marked, but on-site where plumbing modifications are frequently made, information signs are usually placed to alert visitors that recycled water is used on-site. To protect the public water supply, reduced pressure principle backflow prevention devices are commonly required on a domestic water service when recycled water is used on-site—even for irrigation. In dual plumbed buildings, all piping must be labeled, and key shutoff and isolation valves must be tagged with seals that break if the valve is operated. California (and perhaps other states) requires each reuse site to have a trained individual overseeing the recycled water system and operation. Pressure tests and dye tests are periodically required to ensure public health officials that no cross-connection has occurred.

Where a backup water supply system is required, as is commonly the case in dual plumbed buildings, a replaceable pipe spool piece or a "swivel elbow" is normally used on the backup potable water supply or the recycled water system, depending on which system is operational. A reduced pressure principle backflow prevention device is placed on the potable water supply upstream of the removable pipe spool to prevent contamination of the domestic supply. This ensures there is no possibility of both recycled water and potable water entering the recycled water system at the same time.

Typical Recycled Water Notification Sign

Having fire hydrants on recycled water systems can be a cause for concern. Even though they are painted purple, the risk is from the fire truck itself. Pumper trucks typically bring water to fires; if they then connect to a recycled water system, the hoses and water tank need to be thoroughly disinfected before the truck connects to a potable water hydrant. In a large brush fire, where the trucks move about from agency to agency, this may not always be possible. Some agencies have requirements for fire truck disinfection (Tucson, Undated).

To date, direct potable reuse of recycled water is not allowed in the United States. Where recycled water is intended to be indirectly used for domestic water supply—for example, percolation or injection into the ground to blend with the natural groundwater or discharged into a stream that feeds a raw water reservoir—more advanced treatment than that indicated in table 9.1 is normally required. This could involve advanced oxidation processes, reverse osmosis membrane treatment, and high-dose UV disinfection. California has regulations regarding the time recycled water remains underground before being extracted by a potable water well and the volume of recycled water percolated or injected relative to the volume of natural groundwater. Of particular concern are the chemicals of emerging concern.

Landscape and turf irrigation can benefit from the nutrients in the recycled water, particularly nitrogen. Turf grass requires about 0.4 to 1.5 pounds of nitrogen per 1,000 square feet per growing month (Samples & Sorochan, 2008). This is equivalent to 17 to 65 pounds of nitrogen per acre per growing month. Recycled water contains from 16 to 165 pounds of nitrogen per million gallons, depending on the amount of nitrogen required to be removed in the treatment process. The current (2014) price for bulk fertilizer as nitrogen is approximately $0.60 per pound of nitrogen (Farmers Cooperative Association, 2014). So from a fertilizer standpoint, the nitrogen in recycled water is worth 10 to 100 dollars per million gallons.

The use of recycled water to irrigate food crops eaten raw was demonstrated by an eleven-year pilot study in Monterey, California, followed by ten years of actual application on lettuce, broccoli, fennel, strawberries, artichokes, cauliflower, and celery (USEPA, 2012). The recycled water was tertiary (that is, filtered secondary effluent), disinfected to less than 2.2 total coliform per 100 milliliters.

Snowmaking is a good wintertime use of recycled water; however, nutrient reduction (nitrogen and phosphorus) may be necessary to protect streams and lakes when the snow melts and feeds these waters.

Other Water-Quality Requirements

In addition to public health requirements, other water-quality requirements must be met to have a successful recycled water program. For example, some plants are quite sensitive to specific constituents in the water, such as the dissolved mineral content, chloride, sodium, sodium adsorption ratio (SAR), and boron. Depending on the recycled water source and the soil type, some impact may be seen on the plants. It is important to review the water quality with an experienced groundskeeper or golf course superintendent if landscape and turf irrigation is contemplated.

The salinity and composition of recycled water could limit the cycles of concentration in cooling tower reuse. Depending on the quality, it may be beneficial to remove scale-causing constituents.

Contaminants of Emerging Concerns (CECs)

As pointed out above, assisted by new analytical techniques, wastewater professionals are finding that our wastewaters contain many microconstituents not detected in the past. These are antibiotics, pharmaceutical compounds, hormones, and other commonly used drugs such as estrogens, acetaminophen, personal care products such as perfumes and shampoos, antibacterial soap agents, and a host of other compounds. Not only are new compounds showing up in our wastewaters every day as industries continue to develop them, but also their chemistry and composition are unknown because of their proprietary formulation and so are their long-term effects on human health and the environment.

Wastewater industry and scientists are engaged in developing new technologies for the treatment/removal of these microconstituents. These technologies include microfiltration, nanofiltration, reverse osmosis, advanced oxidation such as ultraviolet radiation (UV), ozonation, hydrogen peroxide in combination with UV or ozonation, and activated carbon. Research continues not only to develop new technologies but also to optimize existing technologies to make them more cost-effective.

In spite of all these concerns, there have been no known epidemics attributable to water recycling. More and more communities are engaged in water reuse and recycling, combined with water conservation being practiced by all communities and enterprises with the advent of low-flush toilets, low-water-consuming shower heads, high-efficiency washing machines, and other water-saving devices and fixtures. Also, due to the general awareness of the shortage of water, the per capita water use in major metropolitan centers has constantly come down. Needless to say, the trend will continue.

Implementing a Recycled Water System

Developing a recycled water system seems simple enough, but substantial planning needs to be completed first. A recycled water system is a second water distribution system, complete with storage tanks, pumping stations, transmission and distribution piping, and water services and meters. Constructing this second system is expensive. Ideally, the source of the recycled water (the treatment facility) is in close proximity to the major recycled water users to minimize the piping network. If this is not possible, then consideration should be given to satellite or "skimming" water reclamation plants. These plants are constructed adjacent to major trunk sewers and take, or "skim," water out of the sewer, treat it, and return the residuals (sludges, etc.) to the sewer to be treated at the main treatment plant.

Los Angeles County, the cities of Los Angeles and San Diego, and others have done this on a large scale. One benefit is the sewer now has spare capacity, since a portion of the flow has been removed. That can save the cost of building a parallel sewer. By implementing an advanced wastewater treatment and groundwater recharge project, the Orange County (California) Sanitation District was able to save the cost to expand its ocean outfall. The agency saved money, reduced the environmental impact of a larger ocean discharge flow, and produced a new drinking-water source.

These satellite water-reclamation plants can be constructed to fit in with the community. The use of membrane bioreactor systems allows these facilities to have a relatively small footprint. The membrane bioreactor is a relatively new activated sludge treatment process. The secondary clarification and filtration processes are combined using membranes with small pores immersed in the activated sludge reactor. Pumps draw the treated water through the membranes. A number of these types have been constructed in the basement of buildings for the sole purpose of providing water for toilet and urinal flushing and landscape irrigation.

Another factor that must be considered in the development of a recycled water system is the need for storage, particularly for those that are "irrigation dominated." Unfortunately, the production of wastewater (maximum during the morning and early afternoon) is about twelve hours out of phase with the recycled water demand (typically late night and early morning hours). This results in large storage tanks, pumping stations, and piping systems.

Water recycling plants can blend in with the local area.

Hollow Fiber Membrane

The variation in demands over the year needs to be accommodated. One way would be to provide advanced treatment and groundwater recharge for the off-season flows, with the subsequent extraction to meet peak summertime irrigation demands. A second option is to provide long-term seasonal storage. This may require retreatment of the stored water to remove algae, etc., when the water is pumped to use. Sometimes odors can develop in the recycled water system, particularly during the early years when demands could be low, as pipes were constructed for future demands. The long residence time in the pipes promotes the formation of deposits that extract any remaining chlorine or oxygen out of the water, resulting in a hydrogen sulfide odor when the water is used for spray irrigation.

Summary

This chapter both describes the reasons for recycling water and briefly addresses the water-quality requirements for different uses of recycled water. Although some of the information presented here is based on experiences in California, there is no reason it or its principles cannot be used in other states and communities with or without suitable modifications. The rationale for different standards for different uses of recycled water and the application methods are also provided. By no means is the coverage complete, nor is it intended to be. However, we attempted to describe the general principles and concepts of water reuse so that you can have some understanding of why it is important to recycle water, how water can be recycled, and what risks—if any—there are to humans and the environment in using recycled water as well as some of the considerations in implementing a recycled-water project.

References

Farmers Cooperative Association, I. (2014, May 6). Retrieved May 18, 2014, from http://www.farmersco-op.coop/pages/custom.php?id=21023

Samples, T., & Sorochan, J. (2008). Turfgrass Maintenance. *Developing a Turf Fertilization Plan*. University of Tennessee Extension.

SDCDEH. (2001). Recycled Water Plan Check and Inspection Manual. San Diego County Department of Environmenal Health.

Tucson, C. o. (Undated). Guidelines for Fire Departments. *Procedures for Disinfection of Pumper Trucks & Equipment* . Tucson, AZ, USA: City of Tucson. Retrieved May 18, 2014, from http://www.tucsonaz.gov/water/test/reclaim

USEPA. (2012). *2012 Guidelines for Reuse.* Cincinnati, OH, USA: National Risk Management Research Laboratory.

Chapter 10

Decentralized Systems

Not everyone is close enough to connect to a municipal wastewater system. Residences and commercial establishments in suburban and rural areas must rely on individual on-site systems or small, decentralized wastewater systems. The heart of these systems is the septic tank.

Septic Tanks

Figure 10.1—Typical Septic Tank and Leach Field

A septic tank system consists of a large buried tank of concrete or fiberglass/plastic that collects wastewater from a house or store. Usually the septic tank is in the front or back lawn of a residence or under a parking lot. (See figure 10.1.)

The solids settle to the bottom of the tank, and grease collects on top of the liquid. The tank is always full of wastewater. The effluent from the septic tank flows out of the tank and through a series of perforated pipes under the lawn (called a leach field) and then percolates into the ground. Plants use some of the effluent if their roots penetrate that far.

The solids in a septic tank eventually digest, but not completely. A septic tank needs to be pumped out occasionally by a septic tank pumper truck. The leach fields do not work well if there is groundwater near the surface or if the ground is wet or if there is bedrock near the surface. When this occurs, it is not unusual to see the septic tank effluent rise to the surface and create soggy ground or even a pond. Nitrogen is only partially removed in a septic tank and eventually makes its way into groundwater. Areas with high concentrations of septic tanks usually have high nitrate concentrations in the groundwater.

Septic tanks work well if they are regularly maintained (pumped out) and the subsurface soil profile is favorable. Soils should be permeable, allowing the effluent to percolate readily. If clay or hardpan layers are present, effluent will accumulate and perhaps even come to the surface, creating soggy ground and a health hazard. Remember, the effluent from a septic tank is not disinfected. The system relies on the soil to filter out bacteria and leftover solids. The natural bacteria in the soil remove any remaining organics.

To avoid problems, septic tanks should be inspected about every three to five years, depending on use. Pumping is frequently required at that time. Care should be taken not to dump grease, eggshells, coffee grounds, nonflushable wipes, etc., down the drain. Grease accumulates and can get into the subsurface drainpipes, destroying the leach field and requiring expensive replacement. Eggshells and coffee grounds are not readily biodegradable and just accumulate, taking up valuable space in the tank. Water-softener salt and rinse waters add sodium to the water and may reduce the soil permeability over time.

Where rock or groundwater is close to the ground surface, leach fields cannot be used. Instead, it may be possible to use a mound system or evapotranspiration bed.

An evapotranspiration bed (ET bed) consists of a rubber/plastic membrane or clay-lined excavation about two feet deep. Perforated pipes from the septic tank are surrounded by crushed rock in the bottom of the lined excavation. Fine sand is placed above the crushed rock, and a layer of soil is placed on top of the sand. This allows the plants and turf to grow over the ET bed. The ET bed is typically even with the ground surface, but graded to drain rainfall away. Wastewater from the septic tank is either evaporated or taken up by the overlying plants. Effluent from the septic tank flows by gravity to the ET bed. There is no discharge to the surrounding soils or the water table.

Figure 10.2 Typical Advanced Treatment with Fabric Filters

A mound system is similar to an ET bed, but is constructed above the normal ground surface. It is generally not lined, and effluent from the mound system is allowed to percolate downward. Because the mound system is above the ground surface, the septic tank effluent must be pumped to it.

Where additional nitrogen removal and better effluent quality is desired, advanced treatment processes can follow the conventional septic tank. There are many different types of systems available, including fabric filters (figure 10.2), intermittent sand filters, and passive nitrogen removal systems. These systems reduce the amount of nitrogen discharged to the groundwater. One such system serving a shopping center in Malibu, California, includes nitrogen removal, pressure sand filtration, and ultraviolet disinfection prior to percolation.

After four years of operation, the BOD is less than 5 mg/L, the effluent turbidity is less than 2 NTU, and the total nitrogen in the effluent is less than 3 mg/L.

Cluster Systems

Sometimes septic tank systems can be constructed to serve a cluster of residences. This works well where individual lot sizes are too small to accommodate individual septic tanks and leach fields. Gravity sewers are installed from each residence or wastewater source to a central location where there is sufficient area for the leaching system to be effective. An alternative to a leach field in a clustered system is a seepage pit, which consists of a drilled shaft lined with segmented, perforated concrete pipe sections filled with crushed rock. These can be excavated through tight clay layers that do not allow percolation. Many locations do not allow seepage pits, since they do not achieve nitrogen removal and other water-quality improvements that would otherwise occur in the shallow soil mantel. But additional treatment can mitigate that.

Alternative Collection Systems

Occasionally, individual on-site systems fail over time due to poor soil conditions, maintenance, or design; overuse, or other reasons. Or the number of dwelling units increases to the point that pollution of groundwater or nearby streams and lakes occurs. When this happens, the alternative is often an expensive gravity collection system. There are alternatives to the conventional sewer system.

Septic Tank Effluent Pumping (STEP) Systems

With the STEP system, homeowners keep their septic tanks, assuming they are satisfactory and meet requirements in terms of size, water tightness, etc. A pump is placed in a chamber that pumps the septic tank effluent to a small-diameter pressure sewer in the street. The leach field is disconnected and abandoned. The pressure pipelines in the street can be small-diameter pipes, since the solids have been removed by the septic tank. Furthermore, the pipes can be laid at relatively shallow depths, and since they are under pressure, the pipelines can follow the terrain. In some cases, gravity flow can be maintained, depending on the topography. The shallow, small-diameter pipes make these systems very economical. Monitoring and control systems alert the owner if a pump fails and a high wastewater level in the tank occurs. Wastewater storage is provided in the pump chamber to hold some amount of wastewater in the event of a power outage.

The main sewers in the street lead to a central treatment facility where the effluent can be recycled or disposed of in an environmentally sustainable manner.

Grinder Pump Systems

A grinder pump system is similar to a STEP system, but the septic tank is removed or abandoned. The household wastewater flows to a pump chamber instead, where a pump with chopper blades grinds up solids in the wastewater and pumps them through a shallow, small-diameter pipe to a pipe in the street and on to the treatment facility.

Vacuum Sewer Systems

A vacuum sewer system has vacuum stations throughout the sewer service area that maintain a vacuum in the central sewer system. A small storage tank at each residence accumulates the wastewater. When it is full, a float automatically opens a valve to the vacuum-assisted sewer system, and the contents are sucked out and flow to the vacuum station. At the vacuum station, the sewage is pumped to the treatment facility. As with the STEP and grinder pump systems, vacuum system sewers are laid at shallow depth and can follow the terrain. (See figure 10.3.)

Figure 10.3—Vacuum Sewer System
Courtesy of Airvac, Rochester, Indiana

Decentralized Wastewater Management Districts

Often communities establish decentralized wastewater management districts to oversee the design, installation, monitoring, and maintenance of septic-tank systems and alternative collection systems to ensure continued fail-safe operations. The districts establish design standards and provide periodic inspections, education, training, and regulatory agency reporting. Frequently, this type of organization is required by the regulatory agencies before the implementation of decentralized systems occurs.

Chapter 11

Natural Treatment Systems

Alternatives to highly mechanized, costly, and complex wastewater treatment systems are suitable for small communities. These systems include ponds, lagoons, and other natural treatment systems. These systems are low-cost and easy to operate; they do not require extensive operator training; they have low energy consumption; and they are appropriate if land is available.

Pond Systems

Municipal pond systems can be facultative or aerated, depending on the design and the flows. Industrial ponds are frequently anaerobic. There are estimated to be over eight thousand wastewater treatment ponds, involving more than 50 percent of the wastewater treatment facilities in the United States. Facultative ponds account for 62 percent, aerated ponds for 25 percent, anaerobic for 0.04 percent, and total containment for 12 percent of the pond treatment systems (USEPA, 2011). Total containment ponds have no discharge, but rely on evaporation only.

Facultative Ponds

Figure 11.1 Facultative Ponds at I-70 Rest Area in Kansas

Facultative ponds provide treatment through a symbiotic relationship between algae, which provide oxygen during the day, and bacteria, which use the oxygen produced by the algae and consume the organic matter in the wastewater. (See figure 11.1) Large solids and dead bacteria and algae settle to the bottom of the pond and undergo anaerobic decomposition, releasing methane and hydrogen sulfide. A surface layer of oxygenated water from the algae oxidizes any of the odorous anaerobic decomposition products. Occasionally the ponds get stirred up by wind, which brings up bottom deposits and causes odors. Likewise in colder climates, in spring and fall, the ponds "turn over" due to density differences between the cold water and the warm water, which also causes odors.

Facultative ponds are typically about eight feet deep, are usually arranged in series of two or more ponds, and provide very long retention times—frequently over a month or more. The effluent is relatively low in soluble BOD but will have significant total BOD due to the suspended solids in the effluent. The suspended solids are due to algae, etc. If properly designed and not overloaded, these systems are relatively odor-free.

Mechanically Aerated Ponds

Figure 11.2 Aerated Pond

Frequently, to increase the capacity of facultative ponds, an aeration system is installed to provide more oxygen to the microorganisms and supplement that produced by the algae. The aerators normally float on the surface; occasionally compressed air is added to the bottom of ponds through perforated tubing. (See figure 11.2.) Usually just a small amount of aeration is provided, and the ponds are only partially mixed—that is, most of the solids settle to the bottom and undergo anaerobic decomposition. The aeration is provided to keep a portion of the pond aerobic. Aerated ponds are usually followed by a series of settling ponds.

Increasing the amount of aeration and mixing to the point of being completely mixed can further increase the capacity. Few algae exist in these ponds. These ponds are followed by settling ponds also.

Artificial Wetlands

An artificial wetland consisting of a shallow depression overgrown with bullrushes and reeds is sometimes used to polish the effluent from mechanical secondary treatment facilities or pond systems to provide additional treatment and to remove the algae. These can be either surface or subsurface flow. In the surface flow systems, water at a depth of a few inches flows through the wetlands at a slow rate. The reeds and plants remove nitrogen and suspended solids. (See figure 11.3.)

Figure 11.3 Artificial Wetlands Treatment, Clayton County, Georgia

In subsurface flow systems, the water flows through a permeable layer of sand and small gravel in the bottom. The roots of the plants remove the nutrients.

Figure 11.4 Marsh/Wetlands Treatment System, Arcata, California

Wetlands treatment systems are attractive for small flows, but the land area is extensive for larger facilities.

One of the larger marsh/wetlands system is located in Arcata, California. (See figure 11.4.) The system treats an approximately 2.3 mgd average annual flow. The treatment facility provides primary treatment followed by facultative ponds. The pond effluent flows into a series of primary marshes, then through disinfection, and then into a second set of enhancement marshes. The effluent from the enhancement marshes is disinfected again before discharge into Humbolt Bay in northern California. (Arcata Marsh, 2014)

Reference

Arcata March (2014). Retrieved March 3, 2014 from http://www.ecotippingpoints.org/our-stories/indepth/usa-california-arcata-constructed-wetland-wastewater.html

Chapter 12

The Cost of Wastewater Treatment

Construction Costs

Asking "what is the construction cost of our treatment plant going to be?" is like asking what a residential house will cost. The cost of the house can vary from as low of $50 to as high as $500 per square foot. It depends on a number of variables, such as the type of structure (masonry, concrete, wood, etc.), number of stories, interior amenities, type of floors, size and type of the kitchen and appliances, number of bathrooms, and type and brand of fixtures, to name a few.

The cost of constructing a wastewater treatment plant depends on a number of factors as well, such as

- the capacity of the plant;
- the expansion of an existing plant or a green-field plant (new treatment plant on a new site);
- the processes proposed to treat wastewater and biosolids;
- the discharge and/or water-quality requirements;

- the site characteristics: flat, undulating, sloping, nature of underlying soils, etc.;
- the presence and depth of the groundwater table;
- the cost of the land;
- the cost of other structures and facilities required, such as an operators' building, maintenance building, laboratory, etc.;
- provisions for future expansion;
- the extent of odor control required; and
- the desired outward appearance, such as landscaping.

Based on some or all of these factors, construction costs could vary from ten to twenty-five dollars per gallon a day treated. Thus, the cost per mgd (serving about 10,000 to 12,000 people) could vary from $10 million to $25 million. Looked at another way, the cost allocated to one single-family home generating about 350 to 400 gallons a day of wastewater could vary from $4,000 to $12,000. In a new community with no existing sewers, the cost could be almost two to three times these numbers. The cost could increase further if water reuse and recycling facilities with a water distribution network are provided.

Costs for a sewer system, including the treatment facility, for unsewered communities can be $20,000 to $50,000 per parcel. The cost depends on the topography, density of development, subsurface geology, etc.

If advanced treatment to provide ultrafiltration, reverse osmosis, and advanced oxidation and disinfection are added to permit indirect potable reuse, the cost could range from seven to fifteen dollars per gallon per day, depending on the size of the facility. This would be over and above the cost for secondary treatment.

These are *big* dollar amounts with enormous short-term and long-term funding commitments and associated environmental, socioeconomic, and community impacts. These cannot be taken lightly. Citizens living in affected communities and decision makers responsible for making these commitments on behalf of the citizens they serve should be cognizant of the enormity of these commitments and their impacts on their communities.

The Cost of Operation

Once they are constructed, facilities and plants must be operated and maintained properly. Processes in plants today are quite complex structurally, electrically, mechanically, and process-wise, with sophisticated automation and controls. The day-to-day operation of these plants and scheduled and unscheduled maintenance and repairs require experienced and trained operators, lab technicians, maintenance personnel, and supervisory staff. In addition, operation of these plants requires relatively large amounts of electric power. In smaller treatment facilities, the operating labor costs predominate. All combined, the cost to treat a wastewater flow of 1 million gallons could be anywhere from $1,000 to $3,000, depending on the size of the plant, the processes that comprise it, the effluent quality required, and all the other factors discussed above. Again, these are *big* dollar amounts that must be considered along with the construction cost in the planning of these facilities.

The cost for advanced treatment, as described above, to permit indirect potable reuse would add another $300 to $400 or more per million gallons, again depending on the size of the facility. Larger-capacity facilities can take advantage of the economy of scale.

The cost of maintaining a wastewater collection system is in addition to the cost for treatment. The amount varies from location to location, depending on the age of the collection system, the condition of the pipes, the number of pumping stations, etc.

Cost Reduction Measures

Can the cost of construction and operation be significantly reduced? Unfortunately there are not many ways. At stake here is the health of the community, its overall well-being, and the "happiness index." The judicious planning, design, and operation of these facilities can optimize the construction and operating costs but not reduce them significantly. Following are some ideas for cost optimization and reduction.

- Provide energy-saving equipment such as pumps, blowers, compressors, aeration diffusers, smart lighting, etc. Many utility companies offer monetary incentives if the equipment and processes selected save energy compared to their baseline estimate.

- Incorporate cogeneration facilities in the project: generate electricity with digester gas produced during anaerobic digestion of biosolids and use waste heat for heating and/or production of more electricity.

- Consider producing electric power with solar or wind energy on any spare land or buffer land. Most states, local utility companies, and federal agencies offer monetary assistance for producing solar power.

- Explore the possibilities of securing federal and state grants and low-interest loans to fund these projects. Grants are available, particularly if the project serves economically disadvantaged communities.

- Explore the possibility of obtaining grants if the project involves water reuse. The US Bureau of Reclamation supports and encourages the construction of water reuse and recycling facilities.

- Select a site with favorable topography and subsurface geological and foundation conditions, such as a deep water table, etc.

- Limit provisions for growth and future capacity. Provide capacity only for five to ten years into the future, not twenty to thirty years, since the economy can change and the technologies and the equipment will likely change by the time the future need occurs.

These and other ideas can be helpful in reducing the cost—both construction and operation—of the project. Clearly, this requires good planning, good and prudent design, and smart operation by experienced, knowledgeable, and trained personnel at every single step of the project.

Chapter 13

Sustainability Considerations in Design and Operation

Introduction

As was pointed out in chapter 5, there is now an increased focus on sustainability and carbon-footprint minimization in designing and operating wastewater treatment plants. Future projects must consider the effect of the project on Earth's resources, such as energy, water, and materials, and make every effort to minimize that effect. This makes perfect sense, since it is every generation's responsibility to leave this planet in better shape that it found it. This will be impossible if some of us waste and overuse the limited and ever-diminishing resources by building megahomes, driving gas-guzzling cars, building projects with excessive carbon footprints, not using green and sustainable/renewable energy (such as wind, solar, ocean waves, and hydro), and not imbedding sustainability in our psyche and ethos as we plan and design our projects.

Why the term *carbon footprint*? Energy is required to produce materials, water, electricity—in fact, almost everything. This energy conventionally has been derived in the past from the combustion of carbon-based fuels: diesel, coal, gasoline, wood, etc. The combustion of these fuels produces carbon dioxide—a greenhouse gas (GHG) that is known to trap heat from the sun and thus, scientists believe, contribute to global warming. The more energy a product needs in its manufacturing and transportation, the greater its carbon footprint.

The terms *carbon footprint*, *green*, and *sustainability* should be in the forefront in our planning, design, and operation of all projects. Wastewater plants should be no exception, since they are also consumers of significant quantities of electricity, equipment, and construction materials that require these resources over their lifetime. A wastewater treatment plant that serves 10,000 people, producing about 1 mgd (million gallons per day of wastewater) will require about 100 kilowatts of electricity for secondary treatment. For advanced treatment, the power requirement could be up to 200 kilowatts per mgd, depending on the intended reuse and thus the level of treatment required. With standby equipment and peaking factors included, these numbers could be an additional 50 percent.

Now let us apply these criteria to the entire US population of about 320 million. Power of about 3,200 megawatts will be required just to provide secondary treatment—the minimum treatment required for most reuse applications. When combined with the requirements for treatment of industrial wastes, these numbers will go up significantly, perhaps by a factor of two to three. To put all these numbers in perspective, consider that the output of the largest power plant in the United States, Grand Coulee at the Columbia River in Washington State, is 6,800 megawatts.

Can wastewater treatment plants do their share in reducing this power demand? Of course they can, and they should. Through easy-to-implement, common-sense approaches in planning, design, and operation, plants can save 30 percent or more in power usage and 60 percent or more in in-plant water use for housekeeping, hosing the tanks, seal water for the pumps, landscaping, etc. This all adds up in the sustainability equation, since potable water would otherwise have been needed to meet these essential needs at the plants. Now, if the communities implement area-wide water recycling projects for landscaping and industrial needs, this will only make the projects greener, since the production of potable water in equal amount would have required power in pumping, treatment, and transportation.

What will it take to accomplish the above goals? Not much. Only a change in mind-set and a new ethos that the Earth's resources are limited and must be conserved so that future generations can enjoy these as much, if not more, than we have in our lifetime.

Approaches

There are many approaches—all simple—that can be used to minimize the consumption of resources and the carbon footprint of wastewater projects. They should be thought through and incorporated in all phases of the project: planning, design, construction, and operation. The longer we wait in this process, the less likely it will be that these will be fully incorporated, since each phase of a project affects the following phase. For example, if the design does not incorporate a particular power-saving methodology or equipment (say, a particular type of pump or motor), there is practically no chance that the design will be changed during construction when the contractor is already on board and has provided a bid based on the equipment specified in design. As a result, the project will continue to incur increased expenditure of electric energy over its lifetime in operation.

Therefore, this thinking should happen during the conceptualization of the project.

Planning

Planning is when important issues are brought forth and discussed. It is when the important criteria are formulated and agreed on by the decision makers and stakeholders. Consultants also need to be involved in planning, so that the financial and other impacts of a particular criterion can be factored into the planning process. Take, for example, wastewater treatment plant sizing and location. Should the project be built in stages but expandable to a future size? Should the first phase of the plant consider future discharge requirements, and if so, how will it be upgraded in the future? Should land be acquired for future needs in relation to capacity and discharge requirements?

All these decisions affect carbon footprint, since some additional provisions will be necessary in the first phase of the project—pipes, valves, tank sizes, electric conduits, etc.—with its associated carbon footprint to accommodate these possibilities in the future. If not thought through upfront, future expansions and upgrades may necessitate the decommissioning and/or demolition of some of these facilities. This will result in a twofold carbon footprint caused by both the first phase of construction and the demolition.

Another good example could be the location of the plant. To the extent possible, it should ideally be located at lower elevation than all wastewater contributors so that it does not have to be pumped to where the plant is located. Pumping requires energy: this means adding to the carbon footprint. Another good example is where the plant effluent will be reused. If the plant is far from the users of recycled water, effluent will likely have to be pumped to these users. This means a larger carbon footprint. Again, this can be avoided by advance planning. Could a satellite plant of smaller size be planned and built near the users of recycled water and the larger plant at a different, but lower, elevation? That may be better from the carbon footprint point of view.

So, sustainability and carbon footprint considerations should be made in planning. In fact, that is when they are easier to incorporate into design.

Design

The next step in the implementation of a wastewater treatment project is the predesign and/or design. Typically an elevation difference of about eight to ten feet is required to provide gravity flow between the first and last treatment process within a plant. If filtration and/or membrane processes are needed to meet discharge and/or reuse criteria, greater elevation difference is needed or pumping is required. If the site selected has the necessary slope—or can be regraded to achieve this slope—it can minimize or eliminate the need for pumping. Simply put, wastewater will flow by gravity from one treatment process to the next without intermediate pumping.

If gravity flow cannot be achieved for some reason, there are several ways to reduce the needed elevation difference. Design pipes and channels with low (but realistic) velocities. It is sometimes better to design for a velocity of five to six feet per second at peak flow than a velocity of six to ten feet per second. This may increase the size of the pipe or channel a little, but will reduce the pumping cost and carbon footprint through life of the project. Avoid unnecessary free falling water at tank or pipe outlets. This not only wastes energy but also increases turbulence, which releases odors, increasing the cost of odor control. If pumps are provided (always necessary for several functions within the plant), consider using variable-speed pumps so they perform at optimal efficiency at varying flows received at the plant—both diurnally and seasonally. Use high-efficiency motors in motor-driven pumps.

The list of ways to describe how energy consumption can be reduced by simple means goes on and on. All it requires is a little bit of thinking—but mostly mind-set and awareness. A paper titled "Greener Plants: Designing and Operating a Sustainable Wastewater Treatment Plant," published in the September 2009 issue of the *Water Environment & Technology Journal,* describes this and several other easy-to-implement methods for making the plants green and sustainable.

Besides hydraulics, there are many opportunities in design and operation of wastewater treatment plants where energy savings are possible by simple methods. In a typical modern-day wastewater treatment plant, more than half of total energy/power requirement of an entire plant is for the activated sludge process (a process to oxidize the organic material in wastewater). Blowers or mechanical aerators are used to provide air in this or equivalent processes for oxidation. The efficiency of air/oxygen transfer into sewage (called mixed liquor in this process) ranges from about 5 percent to 20 percent depending on the type of diffusers used to introduce air into the wastewater. Fine-bubble diffusers increase efficiency, while coarse-bubble diffusers cause a reduction of the transfer efficiency. However, fine-bubble diffusers cost more and require cleaning every two to three years to restore their efficiency. But, in the long run, they are cost-effective, save power, and reduce carbon footprint. Energy savings, in most cases, pay for the increased upfront cost in about eight to ten years, depending on the unit cost of power.

Every unit process has a potential for energy savings. Water level drop across filters can range from three to ten feet. Why not consider the use of filters, which result in less water level drop if all other things are equal, such as capital and O&M costs, space requirements, effluent quality, backwash waste as a percent of water filtered, and so on?

Water reuse is another area that enhances sustainability and reduces carbon footprint. Recycle water within the plant as well as outside the plant in highway medians, school playgrounds, city parks and playgrounds, and golf courses, and in industries as cooling water, to name a few. Although groundwater injection generally requires a high level of treatment, it is becoming more feasible as the cost of developing alternative water supplies is continually increasing and will continue to increase as the demand increases, supplies dwindle, and the quality of raw water deteriorates.

Many times, raw water is being imported from large distances. Southern California, for example, is importing a large portion of its raw water from Colorado River and northern California— sources that are two to three hundred miles away. Imagine the cost of importing that water and the associated carbon footprint. Add to this the cost of the power and chemicals to treat it. In addition, these sources are becoming less and less reliable because of protracted drought in the West and an increasing demand of this water by agricultural enterprises and agencies serving municipal and commercial customers—not to mention the concerns of environmentalists. Recycled water, on the other hand, is the most reliable source and has consistent quality.

Another way to achieve overall energy savings at a plant is to produce power using anaerobic digestion (digestion in the absence of air) of biosolids. Many plants in the country are still practicing aerobic digestion of biosolids, which consumes power. Through anaerobic digestion, which produces digester gas—about 60 percent of which is methane—power can be produced. Unlike aerobic digestion, the production of power in anaerobic digestion is approximately 30 to 40 kilowatts from the biosolids generated by 1 mgd of wastewater. Fuel cells, microturbines and internal combustion engines, each with its own pros and cons, can be used to increase the energy recovered from digester gas. This reduces the overall power consumption per mgd from 100 to about 70 kilowatts. With the implementation of other design and operational measures, an overall reduction of 60 percent or more in power use can be easily achieved with a corresponding reduction in carbon footprint.

Buildings at the plant should meet LEED (Leadership in Energy and Environmental Design) certification criteria so that minimum energy is consumed in heating and cooling. The prevailing thought that LEED-certified buildings cost more to construct is somewhat misguided. If planned ahead, they cost no more than non-LEED buildings and consume much less power for heating and cooling. They depend on solar heating and lighting to meet most of their needs and have other green features, which save power.

Think green when designing any building, unit process, or equipment and when selecting materials. This is not a far-fetched and a utopian idea: it is practical and is being increasingly applied in projects. Planners and designers should be aware of this and apply this universally in their projects.

Operations

As when we plan and design, sustainability should remain our focus during the operation of wastewater treatment plants. Efforts by operating personnel to reduce power usage and maximize water reuse should be recognized and rewarded—just as compliance with effluent discharge standards should be. This recognition can be in the form of a certificate given to the operations personnel responsible for the achievement, a monetary bonus or prize, a promotion, or a combination of these.

With these recognitions, energy savings of up to about 80 percent can be achieved, making a plant almost carbon-neutral. Following are possible areas where operations staff can assist in making the operation more sustainable.

- Better monitoring of all parameters indicative of good performance to achieve improved energy efficiency. These include keeping aeration diffusers clean by methods recommended by the manufacturers, ensuring digester gas production at specified levels in the digesters by adequate mixing, heating and controlling other biological parameters such as pH and alkalinity, etc., in the digesters.

- By making sure all equipment is maintained as recommended by the manufacturers so that it performs at maximum efficiency.

- Understanding each unit process and making sure it is operating at its maximum specified efficiency and seeking help if it is not. For example, if digesters are producing poor-quality gas (less than 60 percent methane) and not as much as projected, review operations to see what is causing this. Is it because the biosolids being fed to the digesters are too diluted (say, less than 4 percent solids), which might be a reason for poor digestion? Or is it because the sludge being fed has toxic materials in it—possibly from an industrial dump in the sewers system—that is affecting the quality and the quantity of digester gas?

- By making sure the recycle streams from within the plant (say, from biosolids handling processes) are returned to the front end of the plant during off-peak hours so that they are not overloading the plant during the daytime, when the wastewater flow is high.

- By making sure that the dissolved oxygen level in the aeration basin(s) is not too high (more than 1.5 to 2 mg/L). A high level of oxygen means that the blowers are outputting a larger volume of air than necessary. That means power is being wasted. Adjust the dissolved oxygen setting appropriately so that only the necessary volume of air is discharged into the aeration basins.

- Use plant effluent for in-plant use instead of potable water wherever and whenever possible.

- Understand all equipment and processes at the plant and make sure they are performing at optimum levels.

If they are not, search for the answers and take corrective action immediately.

- Maintain good records of all data on a daily basis and watch for trends. Poor performance and/or a consumption of excessive quantities of chemicals and power do not typically happen suddenly in a biological system. When you monitor trends, corrective steps can be taken immediately rather than days or weeks later, when the solutions become harder to implement.

These are some of the steps operators and maintenance staff can take to improve performance in their plant(s) and operate them at maximum efficiency, not only from a compliance point of view but also from the points of view of conservation of power, chemicals, materials, and water, and thus make the operation more sustainable and ecofriendly.

Other Opportunities for Sustainability

Solar energy. The production of solar energy has taken a big technological leap in recent years. Roof-mounted solar photovoltaic (PV) panels are being installed on roofs of buildings, carports, and garages, and in open spaces wherever possible. A 186-square-meter (2,000-square-foot) home with solar panels installed on its roof can produce enough power for most of the year (five to six kilowatts). This technology can also be used by wastewater treatment plants, which typically have ample roof and open spaces.

With incentives currently offered by utility companies as well as state and federal governments, the production of solar energy has become very affordable and cost-effective. One megawatt of solar energy can be produced in about five acres of land. And most treatment plants have surplus land that can be used for the installation of solar systems. Solar energy produces no greenhouse gases, is ecofriendly, and is sustainable.

Drought-resistant plants. In dry areas, drought-resistant plants can save significant amounts of water, reducing the need for potable and imported water. In addition, these plants are native to the region, thus preserving the ecology and ecosystem of the area. Consideration should also be given to rock planters and groundcover with little or no water need.

Rainwater storage. Rain seldom occurs all year long in most parts of the United States and many other countries. When it pours, much of the rainwater flows to a river or ocean. This excess rainwater can be stored on-site at treatment plants and reused during dry periods. This also will keep pollutants contained in initial surges of storm water from entering our waterways, contaminating them and making them unsightly.

Unpaved parking areas. Some less-used parking areas at a plant can be left unpaved. This will enable rainwater to percolate into the ground and replenish the groundwater aquifer for later withdrawal and reuse.

Green roofs. To further the concept of constructing LEED-certified buildings, consider planting shrubs and flowers on the roof. These keep the interior of the building cool during hot summers, reducing the cost of air-conditioning. In addition, they are attractive and serve as a park for employees and occupants of the building. This idea has merit at plants with large buildings used by operations, design and administrative staff.

Not all the ideas presented here will work for every plant, but some or most can. Just be aware of them, modify and customize them, and build on them.

Save power, water, and other precious resources for future generations. That's simply the right thing to do.

Only the Beginning

These ideas are not all-inclusive but are presented to create awareness and understanding so planners, designers, and operators of wastewater facilities can build on them continually and make their plants more sustainable and ecofriendly. Planners and owners of plants should know what to expect from their designers, and designers should know what to expect from operators. Only by working together can sustainability become not just a goal but also a reality.

Chapter 14

The Future of Wastewater Treatment

We believe the future of wastewater treatment will cover the following areas:

- potable use—both direct and indirect;
- new technologies for nutrient (nitrogen and phosphorus) removal and effluent discharge into sensitive water bodies;
- optimized membrane processes for energy and overall cost reduction;
- removal of recalcitrant and refractory compounds, such as personal care products, hormones, and antibiotics;
- decentralized treatment facilities for optimal water reuse and recycling;
- new technologies for biosolids treatment and reuse;
- new technologies for energy recovery from biosolids, including enhanced gas production with anaerobic digestion; and
- public education.

It should be clear from the above list that the main emphasis in the coming years will be on the reuse of water and therefore the development of new technologies or the optimization of existing technologies to make them more cost-effective and to reduce carbon footprints. The focus will also be on maximizing the reuse of biosolids. To accomplish these objectives, progress will be made both technologically and in public relations. The public will have to be assured that recycled water is safe when properly used. They will have to understand that water is not unlimited and that recycling is necessary and simply the right thing to do. Without it, future generations will have an even more limited water supplies than we currently have, which will adversely affect their living standards. Some even predict that wars between nations will be fought, not for land, but for water.

Doing more with less—that is, enhancing the level of treatment with a reduced consumption of energy, water, and other resources and thus reducing our carbon footprint—will become so imbedded in our culture and way of thinking that projects will be judged based on how much was achieved with few little resources.

With innovations in these and other areas associated with the treatment of wastewater and sludges/biosolids, water and biosolids recycling, combined with increased public awareness and education, treatment plants will essentially be carbon-neutral or nearly so.

Is this a tall order? Not so, we believe.

Glossary

Activated sludge

A secondary wastewater treatment process where the wastewater is mixed with microorganisms in a large tank called an aeration tank. The microorganisms consume the biodegradable organic matter in the wastewater, producing more microorganisms. While in the aeration tank, the microorganism form colonies called floc. The treated wastewater and microorganism floc flow into a secondary clarifier, where the floc settles and separates from the liquid. The liquid, or secondary effluent, is relatively clear with little organic or suspended matter and is ready for disinfection. The settled microorganisms are returned to the aeration tank; some of the microorganisms are removed from the system to maintain a proper balance. Oxygen is provided to the microorganisms in the aeration tank.

Advanced oxidation	An additional treatment process, usually involving ultraviolet irradiation in combination with a strong oxidant like ozone or hydrogen peroxide, which can remove or greatly reduce the concentration of refractory organics and chemicals of emerging concern.
Advanced treatment	A process beyond secondary or tertiary treatment that typically includes membrane treatment with reverse osmosis followed by hydrogen peroxide or a similar strong oxidant in combination with ultraviolet irradiation for disinfection. The purpose of advanced oxidation is to increase the removal of organic compounds. Advanced treatment is normally required if an indirect potable reuse of recycled water is anticipated.
Aerobic	An environment that has some dissolved oxygen in the liquid.
Anaerobic	An environment that does not have any dissolved oxygen in the liquid.
Biosolids	The stabilized solids from a wastewater treatment facility that can be beneficially reused as fertilizer or soil amendment.

BOD	Biochemical Oxygen Demand, measured in milligrams per liter (mg/L), is a measure of the oxygen required by microorganisms to stabilize the biodegradable organics in the wastewater. It is an indicator of the organic strength of the wastewater.
Chemicals of Emerging Concern (CECs)	Chemicals that can be present in water and wastewater in very minute concentrations for which the risk to human health and the environment is unknown. These compounds include antimicrobials, antibiotics, pharmaceuticals and personal care products, and other similar organics. Many of these are not completely removed by conventional wastewater treatment processes.
COD	Chemical Oxygen Demand, measured in mg/L, is a measure of all the chemically oxidizable materials in the wastewater, which includes the biodegradable organics, slowly biodegradable organics, and some small amount of oxidizable inorganic compounds. The COD is always greater than the BOD—almost twice as much in untreated municipal wastewater.

Cogeneration	The process of generating electricity from digester gas and using waste heat for heating and/or producing more electricity. This can be accomplished through a number of different processes
CSO	Combined Sewer Overflow, resulting from the intentional or unintentional discharge of sewage from a combined sewer system into a river, stream, or other water body. This is a problem in many older communities that is being corrected at significant costs.
Detention time or Retention time	The time, usually expressed in hours or days, the wastewater takes to flow through a tank or pond.
Disinfection	The process of killing or inactivating pathogens and other microorganisms so they cannot reproduce or infect humans or animals.

DO	Dissolved Oxygen, measured in mg/L, is a measure of the oxygen concentration that is dissolved in the water or wastewater. Dissolved oxygen is necessary for fish and aquatic life. The microorganisms in the activated sludge process or any other aerobic process depend on dissolved oxygen also. The discharge of incompletely treated wastewater into streams and rivers can cause the dissolved oxygen level in receiving waters to drop to the point where fish cannot survive.
Endocrine disrupting chemicals	Sometimes called "endocrine disruptors." These chemicals may interfere with the body's endocrine system and produce adverse developmental, reproductive, neurological, and immune effects in both humans and wildlife. These chemicals include pharmaceuticals, dioxin and dioxin-like compounds, polychlorinated biphenyls, DDT and other pesticides, and plasticizers such as bisphenol A.

Facultative pond	A wastewater pond that has both an aerobic environment at the surface and an anaerobic environment near the bottom. A facultative pond is commonly used as a wastewater treatment process in small, rural communities where land is plentiful.
Groundwater	Water that occurs below the ground's surface in the pore spaces between soil particles or in rock fractures. Groundwater can be extracted using wells; sometimes groundwater reaches the surface naturally in the form of springs. Approximately 37 percent of water supplied by public water systems is groundwater; in rural areas, almost 100 percent of the water supply is from groundwater.
I/I	Infiltration and inflow is the entrance of groundwater and rainwater into wastewater collection systems through cracks in pipes, leaky joints, openings in manhole covers, and illegal connections. I/I takes up capacity in pipes and treatment facilities that could be used to treat wastewater.

Indicator organism	A microorganism, such as bacteria, that is used as an indicator of the possible presence of pathogens. The presence of the indicator organism is a sign that the water may be contaminated with a pathogen; the absence of the indicator organism provides some degree of certainty that the water is free of pathogen. The most common indicator organism is coliform.
mg/L	Milligrams per liter, a measure of the concentration of a particular substance. It is approximately equal to one part per million as it relates to sewage and wastewater.
NPDES	National Pollutant Discharge Elimination System, the permit process required under the federal Water Pollution Control Act amendments of 1972, which all discharges to surface waters in the United States must obtain before discharging. The permit is typically valid for five years and must be renewed at that time.
Nutrient	Any mineral or substance necessary for growth. It usually refers to nitrogen and phosphorus, which are essential for algal growth. Control of these substances is necessary to control algae.

Pathogen	A microorganism, such as a virus, bacteria, or other form, that causes sickness or death in humans and animals.
pH	A measure of the hydrogen ion concentration in the wastewater. The pH scale extends from 0 to 14. A neutral wastewater has a pH of 7. If it is less than 7, it is acidic; if greater than 7, it is basic, or "caustic." Most municipal wastewaters are between pH 6.5 and 8.
Receiving water	Any water body, such as a stream, river, estuary, lake, ocean, or groundwater, into which treated wastewater is discharged.
Recycled water	Also called "reclaimed water." Wastewater that has been treated to such an extent that it can be reused for a beneficial purpose.
Refractory organic	Organic compounds that are resistant to oxidation by bacteria and other microbial organisms in the secondary treatment process and found in secondary effluent. These compounds can be removed only through advanced treatment.

Septic tank

An underground tank, made of concrete or fiberglass, that provides treatment, settling, and anaerobic digestion of solids, for wastewater generated in rural areas where sewer collection system are not available. The treated wastewater from a septic tank flows into a series of perforated pipes and percolates into the ground. Frequently each house in a rural area where there are no sewers has a septic tank and leach field. The accumulated solids must be pumped out periodically to prevent system failure.

Sludge

The byproduct of wastewater treatment, consisting of the suspended solids that were removed in the primary settling process, the wasted microorganisms and settled material from the secondary treatment process, and chemical and organic solids from tertiary treatment processes. Sludge does not become biosolids until it has been stabilized through digestion or similar processes and can be put to beneficial reuse.

SSO	Sanitary Sewer Overflow, the unintentional discharge of sewage from a separate sanitary sewer system into a river, stream, or other water body. These are strictly prohibited and can result in substantial fines and penalties.
Stabilization	The process of converting biodegradable sludge solids to methane and carbon dioxide (anaerobic digestion) or carbon dioxide and water (aerobic digestion). This makes the solids less odorous.
Sustainability	In broad terms, sustainability means meeting the needs of the present without compromising the ability of future generations to meet their needs. In the present context, it means involving methods that do not use up or destroy natural resources such as water, energy, and other materials.

TDS	Total Dissolved Solids, measured in mg/L, is a measure of the dissolved minerals in wastewater, such as sodium, calcium, magnesium, chloride, sulfate, bicarbonate, and other ions. This is what contributes to salinity in wastewater and can be a concern when the wastewater is recycled and used for agricultural or landscape irrigation. The TDS concentration in wastewater is always greater than the TDS in a drinking water supply. The TDS is determined by evaporating a known volume of wastewater, which has passed through a filter, and determining the weight of the minerals left behind.
TMDL	Total Maximum Daily Load, the maximum amount of a contaminant, usually expressed as pounds/day or kilograms/day, that a receiving water is able to accommodate without adversely impacting its beneficial use or violating its water-quality standard.
Trickling filter/ biotower	A treatment process where the wastewater is sprinkled over a bed of rocks or a plastic media containing a film of microorganisms. Oxygen is provided either through natural ventilation or by large fans.

TSS — Total Suspended Solids, measured in mg/L, is a measure of the suspended material in wastewater. It is determined by filtering the wastewater through a glass fiber filter, drying the filter and captured solids, and determining the difference in weight between the filter and the dried solids.

µg/L — Microgram per liter, a measure of the concentration of a substance. It is approximately equal to one part per billion. It is 1/1000 of a mg/L.

Unit Conversions

1 gallon = 3.785 Liters (L)

1 ft^3 = 7.48 gallons

1 lb = 0.454 kilograms (kg)

1 lb = 454 grams (g)

1 m^3 = 264 gallons

1 m^3 = 35.29 ft^3

1 million gallons per day (1 mgd) = 3785 m^3/day

1 million gallons per day (1 mgd) = 157.7 m^3/hour

1 million gallons per day (1 mgd) = 43.8 L/second

1 acre = 43,560 ft^2

1 hectare = 2.47 acres

1 ton = 2000 lb

1 metric tonne = 1000 kg = 2200 lb

1 BTU = 252 calories

1 British Thermal Unit (BTU) = 1055 Joules (J)

1 BTU/ft^3 = 37.26 kiloJoules (kJ)/m^3

1 mg/L = 1 part per million (ppm), (approximately)

Made in the USA
Middletown, DE
18 January 2022

59014024R00104